PHYSICAL AND CHEMICAL PROPERTIES OF CARBON NANOTUBES

PHYSICAL AND CHEMICAL PROPERTIES OF CARBON NANOTUBES

Editor

Hua Fen Han

scitus
academics

Physical and Chemical Properties of Carbon Nanotubes
Edited by **Hua Fen Han**

ISBN: 978-1-68117-669-7
Library of Congress Control Number: 2016946535

SCITUS Academics LLC,
616, Corporate Way, Suite 2, 4766,
Valley Cottage, NY 10989

www.scitusacademics.com

Notice

Printed in United States of America on Acid Free Paper

PREFACE

A great deal of research has been devoted to the study of carbon nanotubes' balance structure. In other words, carbon nanotubes can be found as a carbon macro molecule, analogous to a graphite sheet rolling into a cylinder. Carbon nanotubes are rolled up graphene sheets with a nanometer scale diameter quasi-one-dimensional structure. Due to its excellent structural, optical, electronic, chemical and physical properties, carbon nanotubes have attracted a great deal of attention from physicists, chemists, material scientists and electronic device engineers in the last 20 years.

This compilation covers recent issues related to the physical, structural, electrical and chemical properties of carbon nanotubes. There are many applications for carbon nanotubes that use CNT's unique aspect ratio, physical strength, and thermal and electrical conductivity properties. In addition, carbon nanotubes directly benefit from a rich history and a large knowledge of carbon chemistry. Therefore, carbon nanotubes can be functionalized in many ways using a variety of well-known chemical reactions. In addition, the geometry of a nanotube makes it impossible to develop new synthetic structures along with other carbon structures. Carbon nanotube research, especially for industrial applications, is becoming more important.

TABLE OF CONTENTS

Chapter 1 Elastic Properties of Carbon Nanotubes 1

Chapter 2 Nickel supported on nitrogen-doped carbon nanotubes as hydrogen oxidation reaction catalyst in alkaline electrolyte 39

Chapter 3 Characterization of Laser-Induced Defects and Modification in Carbon Nanotubes by Raman Spectroscopy 61

Chapter 4 Growth of semiconducting single-wall carbon nanotubes with a narrow band-gap distribution 93

Chapter 5 Vibroelectronic Properties of Functionalized Single-Walled Carbon Nanotubes and Double-walled Boron Nitride Nanotubes 115

Chapter 6 Efficient near-infrared up-conversion photoluminescence in carbon nanotubes 169

Chapter 7 The necessary length of carbon nanotubes required to optimize solar cells 187

Chapter 8 General synthesis of complex nanotubes by gradient electrospinning and controlled pyrolysis 199

Chapter 9 Microwave Dielectric Properties of Carbon Nanotube Composites 223

Chapter 10 Quantum-coupled radial-breathing oscillations in double-walled carbon nanotubes 249

Chapter 11 Carbon Nanotubes Produced from Ambient Carbon Dioxide for Environmentally Sustainable Lithium-Ion and Sodium-Ion Battery Anodes 267

Index 285

1

ELASTIC PROPERTIES OF CARBON NANOTUBES

Qiang Han[1] and Hao Xin

[1] School of Civil Engineering and Transportation, South China University of TechnologyGuangzhou, People's Republic of China

1. INTRODUCTION

The geometric structure of carbon nanotubes (CNTs) can be considered to be the curling of graphene (graphite sheet) (Gao & Li, 2003 ; Shen, 2004). The physical parameters of carbon in graphite are widely adopted in the molecule dynamics (MD) simulations or other theoretical studies of CNTs, such as the bond energy and length of C-C, the bond angle of C-C-C, etc (Belytschko et al., 2002; Mayo et al., 1990; Xin et al., 2007, 2008). Previous researches have shown that the elastic modulus of small CNTs changes a lot when the radius varies. However, the elastic modulus of CNTs fairly approaches that of graphene, while the radius of CNTs is large enough. Therefore, it is necessary to have a clear understanding of the mechanical properties of graphene for the further realization of the properties of CNTs.

Numerous researchers carried out experiments to measure the effective elastic modulus of CNTs (Krishnan et al., 1998; Poncharal et al., 1999). They reported the effective Young's modulus of CNTs ranging from 0.1 to 1.7 TPa, decreasing as the diameter increased, and the average was about 1.0~1.2 TPa. MD simulations have provided abundant results for the understanding of the buckling behavior of CNTs. The Young's modulus of the CNTs was predicted about 1.0~1.2 TPa through various MD methods (Jin & Yuan, 2003; Li & Chou, 2003; Lu, 1997). Hu et al. (Hu et al., 2007) proposed an improved molecular structural mechanics method for the buckling analysis of CNTs, based on Li and Chou's model (Hwang et al., 2010) and Tersoff-Brenner potential (Brenner, 1990).

Due to the different methods employed on various CNTs in these researches, the reported data scattered around an average of 1.0 TPa.

The elastic properties were also discussed in the theoretical analysis by Govinjee and Sackman (Govinjee and Sackman, 1999) based on Euler beam theory, which showed the size dependency of the elastic properties at the nanoscale, which does not occur at continuum scale. Harik (Harik, 2001) further proposed three non-dimensional parameters to validate the beam assumption, and the results showed that the beam model is only proper for CNTs with small radius. Liu et al (Liu et al., 2001) reported the decrease of the elastic modulus of CNTs with increase in the tube diameter. Shell model was also used in some researches (Wang et al., 2003), to study axially compressed buckling of multi-walled CNTs. And studies by Sudak (Sudak, 2003) reported that the scale effect of CNTs should not be ignored. Wang et al (Wang et al., 2006) investigated the buckling of CNTs and the results showed that the critical buckling load drawn with the classical continuum theory is higher than that with considering the scale effect. There are some other researches also reporting clear scale effect on the vibration of CNTs (Wang & Varadan, 2007; Zhang et al., 2005). Li and Chou (Li & Chou, 2003) put forward a truss model for CNTs and their studies showed the radius-dependence of elastic modulus of SWCNTs. Finite element analysis (FEA) is also employed in researches on the mechanical properties of CNTs (Yao & Han, 2007, 2008 ; Yao et al., 2008).

An equivalent model is established in this chapter, based on the basic principles of the anisotropic elasticity and composite mechanics, for the analysis of the elastic properties of graphite sheet at the nanometer scale. With this equivalent model, the relationship between the nanotube structure and the graphite sheet is built up, and the radial scale effect of the elastic properties of CNTs is investigated.

2. CONSTITUTIVE EQUATIONS OF ORTHOTROPIC SYSTEM

The essential difference between the basic equations of anisotropic materials and those of isotropic ones is in their constitutive equations, which means the usage of the anisotropy Hooke law for the anisotropic constitutive equation and the isotropy one for the other. The phenomenon reflected from the anisotropy equations is more accurate than from the isotropy ones, though this distinction also makes the calculation with the anisotropy equations much more complicated.

One of the anisotropic systems, with three mutually perpendicular principal axes of elasticity, is called an orthotropic system (Fig. 1). If the three principal axes of elasticity are defined as x_1, x_2 and x_3, the constitutive equation of orthotropic system can be obtained as follow,

$$\begin{Bmatrix} \varepsilon_1 \\ \varepsilon_2 \\ \varepsilon_3 \\ \varepsilon_4 \\ \varepsilon_5 \\ \varepsilon_6 \end{Bmatrix} = \begin{bmatrix} S_{11} & S_{12} & S_{13} & & & \\ S_{12} & S_{22} & S_{23} & & & \\ S_{13} & S_{23} & S_{33} & & & \\ & & & S_{44} & & \\ & & & & S_{55} & \\ & & & & & S_{66} \end{bmatrix} \begin{Bmatrix} \sigma_1 \\ \sigma_2 \\ \sigma_3 \\ \sigma_4 \\ \sigma_5 \\ \sigma_6 \end{Bmatrix} \tag{1}$$

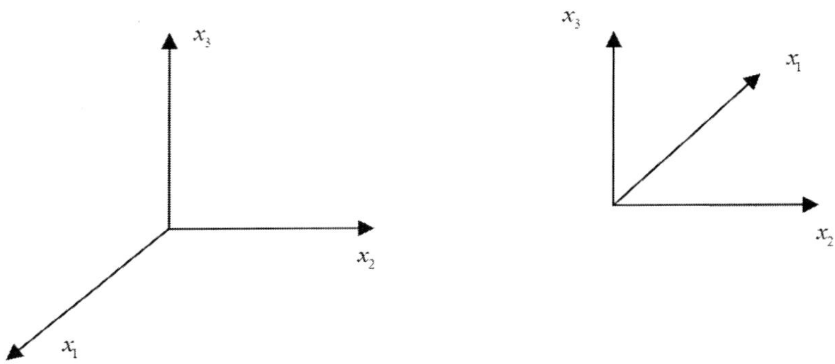

Figure 1. The principal elastic axes of orthotropic system: a) right-hand coordinate system; b) left-hand coordinate system

And the inverse function of the equation (1) is,

$$\begin{Bmatrix} \sigma_1 \\ \sigma_2 \\ \sigma_3 \\ \tau_{23} \\ \tau_{31} \\ \tau_{12} \end{Bmatrix} = \begin{bmatrix} C_{11} & C_{12} & C_{13} & & & \\ C_{12} & C_{22} & C_{23} & & & \\ C_{13} & C_{23} & C_{33} & & & \\ & & & C_{44} & & \\ & & & & C_{55} & \\ & & & & & C_{66} \end{bmatrix} \begin{Bmatrix} \varepsilon_1 \\ \varepsilon_2 \\ \varepsilon_3 \\ \gamma_{23} \\ \gamma_{31} \\ \gamma_{12} \end{Bmatrix} \tag{2}$$

where,

$$\begin{cases} S_{\substack{11\\22\\33}} = (C_{\substack{22\\33\\11}} C_{\substack{33\\11\\22}} - C^2_{\substack{23\\31\\12}}) / C \\ S_{\substack{23\\31\\12}} = (C_{\substack{21\\32\\13}} C_{\substack{31\\12\\23}} - C_{\substack{23\\31\\12}} C_{\substack{11\\22\\33}}) / C \\ S_{\substack{44\\55\\66}} = 1 / C_{\substack{44\\55\\66}} \\ C = C_{11}C_{22}C_{33} - C_{11}C_{23}^2 - C_{22}C_{31}^2 - C_{33}C_{12}^2 + 2C_{12}C_{23}C_{31} \end{cases} \tag{3}$$

Equation (3) is also correct if we make exchanges of *C* for *S* and *S* for *C*.

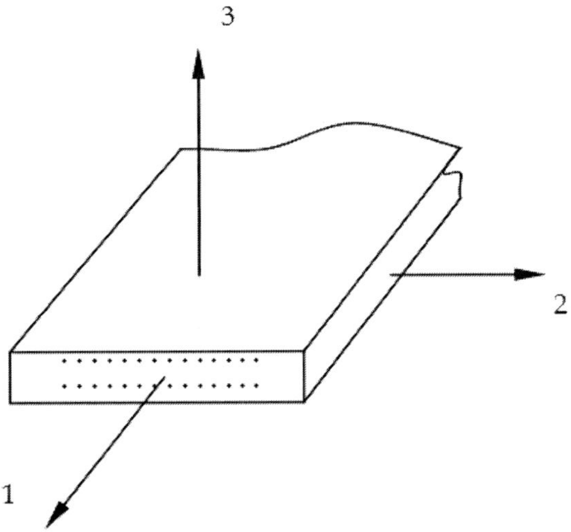

Figure 2. Sketch of macroscopic homogeneous orthotropic material

The unidirectional fiber composed composite materials can be treated as orthotropic systems, for which the three axes in the right-handed coordinate system in Fig. 2 are the principal material axes and the axis 1 is along the fiber length. Thus, the components of equation (1) can be given minutely in detail,

$$\varepsilon_1 = S_{11}\sigma_1 + S_{12}\sigma_2 + S_{13}\sigma_3 = \frac{1}{E_1}\sigma_1 - \frac{\nu'_{12}}{E_2}\sigma_2 - \frac{\nu'_{13}}{E_3}\sigma_3$$

$$\varepsilon_2 = S_{21}\sigma_1 + S_{22}\sigma_2 + S_{23}\sigma_3 = -\frac{\nu'_{21}}{E_1}\sigma_1 + \frac{1}{E_2}\sigma_2 - \frac{\nu'_{23}}{E_3}\sigma_3$$

$$\varepsilon_3 = S_{31}\sigma_1 + S_{32}\sigma_2 + S_{33}\sigma_3 = -\frac{\nu_{31}}{E_1}\sigma_1 - \frac{\nu_{32}}{E_2}\sigma_2 + \frac{1}{E_3}\sigma_3$$

$$\gamma_{23} = S_{44}\tau_{23} = \frac{1}{G_{23}}\tau_{23}$$

$$\gamma_{31} = S_{55}\tau_{31} = \frac{1}{G_{31}}\tau_{31}$$

$$\gamma_{12} = S_{66}\tau_{12} = \frac{1}{G_{12}}\tau_{12} \tag{4}$$

where,

$$\begin{cases} S_{ij} = S_{ji} \\ \dfrac{\nu_{ij}}{E_j} = \dfrac{\nu_{ji}}{E_i} \end{cases} \tag{5}$$

3. MACRO-MECHANICS FUNDAMENTAL PRINCIPLE OF COMPOSITE MATERIAL

3.1. Constitutive Equations Of Monolayer Under Plane Stress

The monolayer composite with unidirectional fiber can be considered as a homogeneous orthotropic material in the macro analysis. Fig. 3 displays the three principal material axes and the axis 3 is perpendicular to the mid-plane of the monolayer. Suppose the monolayer is in a plane stress state, there are the in-plane stresses,

$$\sigma_1, \quad \sigma_2, \quad \tau_{12}(\sigma_6) \tag{6}$$

and the out-plane stresses,

$$\sigma_3 = \tau_{23}(\sigma_4) = \tau_{13}(\sigma_5) = 0 \tag{7}$$

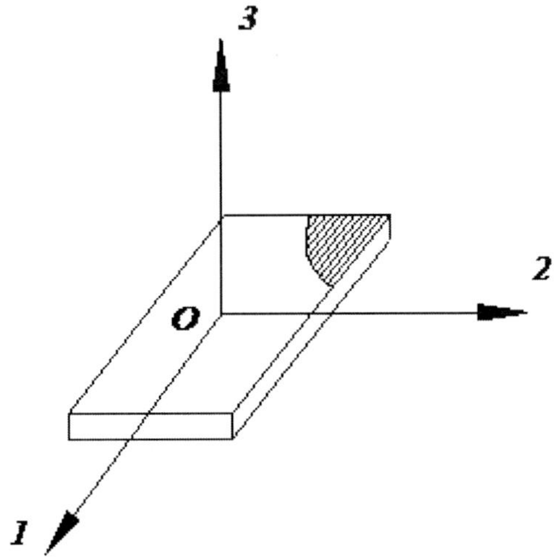

Figure 3. Sketch of monolayer composite with unidirectional fiber

Thus, the constitutive equation (1) can be given in two parts as follows,

$$\varepsilon_3 = S_{31}\sigma_1 + S_{32}\sigma_2 = -\frac{\nu_{31}}{E_1}\sigma_1 - \frac{\nu_{32}}{E_2}\sigma_2$$

$$\gamma_{23} = 0, \quad \gamma_{31} = 0 \tag{8}$$

$$\varepsilon_1 = S_{11}\sigma_1 + S_{12}\sigma_2 = \frac{1}{E_1}\sigma_1 - \frac{\nu_{12}}{E_2}\sigma_2$$

$$\varepsilon_2 = S_{21}\sigma_1 + S_{22}\sigma_2 = -\frac{\nu_{21}}{E_1}\sigma_1 + \frac{1}{E_2}\sigma_2$$

$$\gamma_{12} = S_{66}\tau_{12} = \frac{1}{G_{12}}\tau_{12} \tag{9}$$

The equation (8) is for the out-plane strain and the equation (9) for the in-plane strain. And the equation (9) can also be written in a matrix form,

$$\begin{Bmatrix} \varepsilon_1 \\ \varepsilon_2 \\ \gamma_{12} \end{Bmatrix} = \begin{bmatrix} S_{11} & S_{12} & 0 \\ S_{21} & S_{22} & 0 \\ 0 & 0 & S_{66} \end{bmatrix} \begin{Bmatrix} \sigma_1 \\ \sigma_2 \\ \tau_{12} \end{Bmatrix} \text{ or } \{\varepsilon^1\} = [S]\{\sigma^1\} \tag{10}$$

where [S] is the axis flexibility matrix. The equation (10) can also be given as,

$$\begin{Bmatrix} \varepsilon_1 \\ \varepsilon_2 \\ \gamma_{12} \end{Bmatrix} = \begin{bmatrix} \frac{1}{E_1} & -\frac{\nu_{12}}{E_2} & 0 \\ -\frac{\nu_{21}}{E_1} & \frac{1}{E_2} & 0 \\ 0 & 0 & \frac{1}{G_{12}} \end{bmatrix} \begin{Bmatrix} \sigma_1 \\ \sigma_2 \\ \tau_{12} \end{Bmatrix} \tag{11}$$

The inverse of the equation (10) is available,

$$\begin{Bmatrix} \sigma_1 \\ \sigma_2 \\ \tau_{12} \end{Bmatrix} = \begin{bmatrix} Q_{11} & Q_{12} & 0 \\ Q_{21} & Q_{22} & 0 \\ 0 & 0 & Q_{66} \end{bmatrix} \begin{Bmatrix} \varepsilon_1 \\ \varepsilon_2 \\ \gamma_{12} \end{Bmatrix} \text{ or } \{\sigma^1\} = [Q]\{\varepsilon^1\} \tag{12}$$

where [Q] is the converted axis stiffness matrix at the plane stress state. Qij in equation (12) and Cij in equation (2) have the following relations,

$$Q_{\substack{11\\12\\22}} = C_{\substack{11\\12\\22}} - C_{\substack{13\\13\\23}}C_{\substack{13\\23\\23}} / C_{33}$$

$$Q_{66} = C_{66} \tag{13}$$

The values of the flexibility and stiffness in equation (11) and equation (13) can be obtained through micromechanics calculations or experiments.

3.2. Off-Axis Flexibility and Stiffness Of Monolayer Under Plane Stress

The off-axis flexibility and stiffness are often used in the mechanical analysis on the unidirectional fiber monolayer. As displayed in Fig. 4, axes 1 and 2 are along the principal axes of material, x and y are off-axis, and the anticlockwise angle from x-axis to the 1-axis is positive. Thus, we obtain,

$$\begin{Bmatrix} \sigma_1 \\ \sigma_2 \\ \tau_{12} \end{Bmatrix}^x = \begin{bmatrix} \cos^2\theta & \sin^2\theta & -2\sin\theta\cos\theta \\ \sin^2\theta & \cos^2\theta & 2\sin\theta\cos\theta \\ \sin\theta\cos\theta & -\sin\theta\cos\theta & \cos^2\theta-\sin^2\theta \end{bmatrix} \begin{Bmatrix} \sigma_1 \\ \sigma_2 \\ \tau_{12} \end{Bmatrix}^1$$

$$\text{or} \quad \{\sigma^x\} = [T]\{\sigma^1\} \tag{14}$$

and

$$\begin{Bmatrix} \sigma_1 \\ \sigma_2 \\ \tau_{12} \end{Bmatrix}^1 = \begin{bmatrix} \cos^2\theta & \sin^2\theta & 2\sin\theta\cos\theta \\ \sin^2\theta & \cos^2\theta & -2\sin\theta\cos\theta \\ -\sin\theta\cos\theta & \sin\theta\cos\theta & \cos^2\theta-\sin^2\theta \end{bmatrix} \begin{Bmatrix} \sigma_1 \\ \sigma_2 \\ \tau_{12} \end{Bmatrix}^x$$

$$\text{or} \quad \{\sigma^1\} = [T]^{-1}\{\sigma^x\} \tag{15}$$

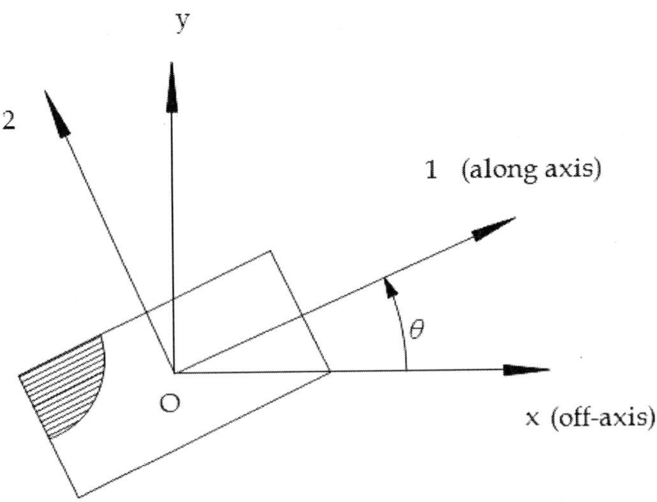

Figure 4. Sketch of coordinate transformation between the axis and off-axis of monolayer with unidirectional fibers

If, we set,

$$[R] = \begin{bmatrix} 1 & & \\ & 1 & \\ & & 2 \end{bmatrix} \tag{16}$$

$$[F] = [R][T][R]^{-1} \begin{bmatrix} \cos^2\theta & \sin^2\theta & -\sin\theta\cos\theta \\ \sin^2\theta & \cos^2\theta & \sin\theta\cos\theta \\ 2\sin\theta\cos\theta & -2\sin\theta\cos\theta & \cos^2\theta - \sin^2\theta \end{bmatrix} \tag{17}$$

the follow can be obtained,

$$\{\varepsilon^x\} = [F]\{\varepsilon^1\} \tag{18}$$

It should be known from the equation (15) and (17) that,

$$[F]^T = [T]^{-1} \tag{19}$$

So that,

$$[F]^{-1} = [T]^T \tag{20}$$

Thus, the inverse function of the equation (18) is,

$$\{\varepsilon^1\} = [F]^{-1}\{\varepsilon^x\} = [T]^T\{\varepsilon^x\} \tag{21}$$

With these relations, the follows can be given,

$$\{\sigma^x\} = [T]\{\sigma^1\} = [T][Q]\{\varepsilon^1\} = [T][Q][T]^T\{\varepsilon^x\}$$
$$\text{or } \{\sigma^x\} = [\bar{Q}]\{\varepsilon^x\} \tag{22}$$

and

$$\{\varepsilon^x\} = [F]\{\varepsilon^1\} = [F][S]\{\sigma^1\} = [F][S][T]^{-1}\{\sigma^x\} = [F][S][F]^T\{\sigma^x\}$$
$$\text{or } \{\varepsilon^x\} = [\bar{S}]\{\sigma^x\} \tag{23}$$

The equation (22) and (23) are the constitutive equations in the Oxy coordinate system, where the off-axis stiffness $[\bar{S}]$ and the off-axis flexibility $[\bar{Q}]$ are given as follows,

$$[\bar{Q}] = \begin{bmatrix} \bar{Q}_{11} & \bar{Q}_{12} & \bar{Q}_{16} \\ \bar{Q}_{21} & \bar{Q}_{22} & \bar{Q}_{26} \\ \bar{Q}_{61} & \bar{Q}_{62} & \bar{Q}_{66} \end{bmatrix} = [T][Q][T]^T \tag{24}$$

$$[\bar{S}] = \begin{bmatrix} \bar{S}_{11} & \bar{S}_{12} & \bar{S}_{16} \\ \bar{S}_{21} & \bar{S}_{22} & \bar{S}_{26} \\ \bar{S}_{61} & \bar{S}_{62} & \bar{S}_{66} \end{bmatrix} == [F][S][F]^T \tag{25}$$

In the above two equations,

$$\bar{Q}_{ij} = \bar{Q}_{ji}, \quad \bar{S}_{ij} = \bar{S}_{ji} \tag{26}$$

The off-axis stiffness is,

$$\begin{aligned}
\bar{Q}_{\substack{11\\22}} &= Q_{\substack{11\\22}} \cos^4\theta + 2(Q_{12} + 2Q_{66})\sin^2\theta\cos^2\theta + Q_{\substack{22\\11}} \sin^4\theta \\
\bar{Q}_{\substack{12\\66}} &= Q_{\substack{12\\66}} + (Q_{11} + Q_{22} - 2Q_{12} - 4Q_{66})\sin^2\theta\cos^2\theta \\
\bar{Q}_{\substack{16\\26}} &= \pm(Q_{\substack{11\\22}} - Q_{12} - 2Q_{66})\sin\theta\cos^3\theta \mp (Q_{\substack{22\\11}} - Q_{12} - 2Q_{66})\sin^3\theta\cos\theta
\end{aligned} \tag{27}$$

And the off-axis flexibility,

$$\begin{aligned}
\bar{S}_{\substack{11\\22}} &= S_{\substack{11\\22}} \cos^4\theta + (2S_{12} + S_{66})\sin^2\theta\cos^2\theta + S_{\substack{22\\11}} \sin^4\theta \\
\bar{S}_{12} &= S_{12} + (S_{11} + S_{22} - 2S_{12} - S_{66})\sin^2\theta\cos^2\theta \\
\bar{S}_{66} &= S_{66} + 4(S_{11} + S_{22} - 2S_{12} - S_{66})\sin^2\theta\cos^2\theta \\
\bar{S}_{\substack{16\\26}} &= \pm(2S_{\substack{11\\22}} - 2S_{12} - S_{66})\sin\theta\cos^3\theta \mp (2S_{\substack{22\\11}} - 2S_{12} - S_{66})\sin^3\theta\cos\theta
\end{aligned} \tag{28}$$

3.3. Constitutive Equations In Classical Laminated Plate Theory

The so-called classical laminated plate theory or the classical laminate theory, refers to the use of the straight normal hypothesis in elastic shell theory, neglects a number of secondary factors, and has been an acknowledged laminated plate theory. In the classical laminated plate theory, the transverse shear strain $\gamma 23$ and $\gamma 31$, and the normal direction strain $\varepsilon 3$ are supposed to be zero.

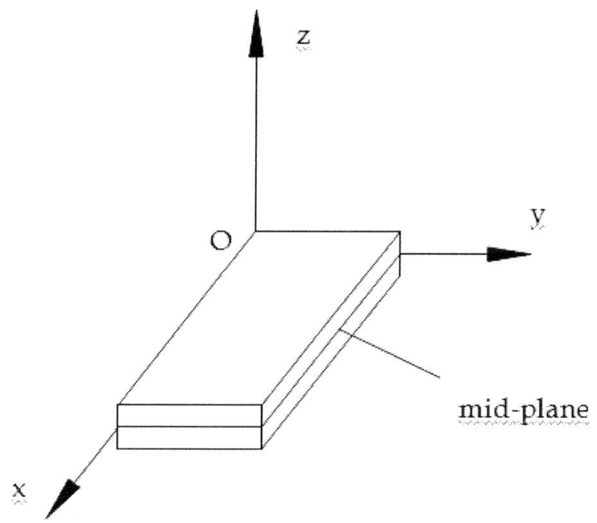

Figure 5. Sketch of laminated thin plate and the Cartesian coordinates

As a result of straight normal assumption, the deformation of laminated plate can be described with the mid-plane deformation. If the mid-plane strain is $\varepsilon 0$ and thickness of each layer in the laminated plate is the same, the mid-plane stress of an n-layers laminated plate is

$$\left\{\sigma^0\right\} = \frac{1}{n}\sum_{k=1}^{n}\left\{\sigma^{(k)}\right\} = \frac{1}{n}\sum_{k=1}^{n}\left(\left[\bar{Q}\right]^{(k)}\left\{\varepsilon^0\right\}\right) = \left[\frac{1}{n}\sum_{k=1}^{n}\left(\left[\bar{Q}\right]^{(k)}\right)\right]\cdot\left\{\varepsilon^0\right\} \tag{29}$$

where σ0 is the mid-plane stress of the laminated plate, σ(k) is the stress of the *k*th layer, $[Q]^{(k)}$ is the converted stiffness of the *k*th layer.

4. EQUIVALENT MODEL OF GRAPHITE SHEET

4.1. The Basic Idea Of The Equivalent Model

All the C atoms in the graphite sheet are connected with the σ bonds and the bonds form a hexagonal structure (Fig.6). Fig. 7 is a schematic diagram of the graphite sheet, thick solid lines in which represent the C-C bond in graphite. If each C-C bond is longer (shown in thin lines), that will form a network structure, as shown in Fig. 8.

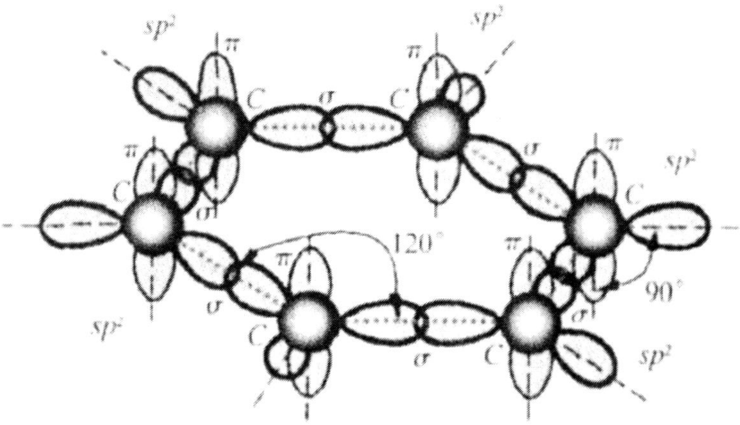

Figure 6. The bonding relationship in C-C covalent bonds

As can be seen from Fig. 8, the network structure is formed by the three groups of parallel fibers and they are into 60 degree angles with each other. If we consider each group of fiber as a composite monolayer, the mechanical properties of the entire network structure can be obtained with the laminated plate theory. Comparing Fig. 7 and Fig. 8, we find that the network structure in Fig. 8 can also be formed if the three graphite sheets in Fig. 7 are staggered and stacked one on top of the other.

In summary, here we put forward a new original equivalent model used to study the mechanical properties of graphite sheet. The analysis steps are, treat the network structure shown in Fig. 8 as a laminated composite plate with three layers orthotropic monolayer of unidirectional fiber (each fiber in the monolayer is just the covalent C-C bond in series), the mechanical properties of fiber can be deduced from the physical parameters of graphite, and the 1/3 of the converted stiffness of the network structure can be considered as the converted stiffness of the graphite sheet at the plane stress state.

Figure 7. The atomic structure of a graphite sheet

4.2 Mechanical Properties Of Graphite Sheet At Plane Stress State

The hexagonal plane composed of the σ bonds is defined as the σ-plane, and the energy of interactions between any C atoms in the σ-plane are considered to be functions of the position of the C atoms. With all the weak interactions (e.g. the electronic potential, the van der Waals interactions) neglected, the total potential energy of the graphite sheet can be expressed as.

$$U_{graphite} = U_r + U_\theta \tag{30}$$

where U_r is the axial stretching energy of C-C bond, U_θ is the C-C-C bond angle potential. Establish a local coordinate system in the σ-plane with the C-C bond direction as the x′ axis (Fig. 9), we can obtain the whole potential,

$$U_{C-C} = U_{x'} + U_{\theta'} \tag{31}$$

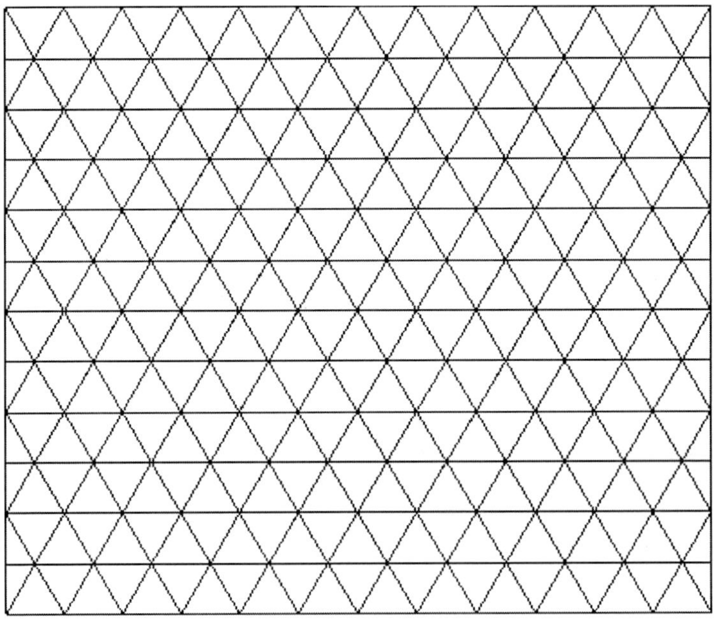

Figure 8. Effective network structure of laminated graphite sheets

where Ux′ is the axial stretching energy of C-C bond in the local coordinate system and Uθ′ the angle bending energy. According to the basic principles of molecular mechanics, the two bond energy can be written as:.

$$U_{x'} = \frac{1}{2} k_{x'} \cdot \delta_{x'}^2$$
$$U_{\theta'} = \frac{1}{2} k_{\theta'} \cdot \delta_{\theta'}^2 \qquad (32)$$

where δx′ and δθ′ are the displacement, kx′ and kθ′ are the MD force field constants.

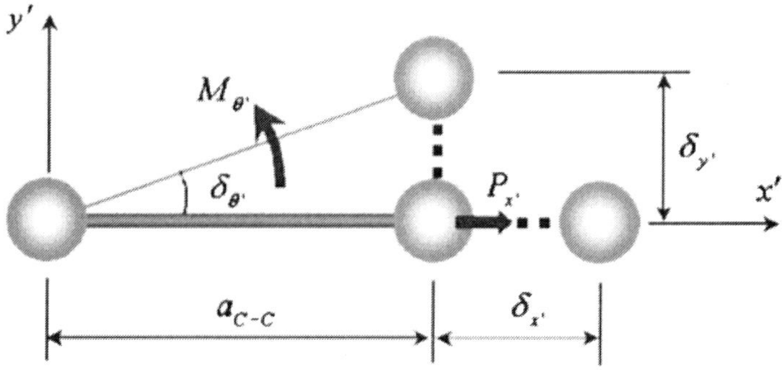

Figure 9. Local element coordinates of C-C bond in graphite

4.3 Elastic constants of monolayer in the equivalent model

If we consider the C-C bond as a single elastic fiber in the equivalent model, and the 1 axis for the fiber is defined as the axial direction of the C-C bond, we have,

$$K_1 = k_{x'} \qquad (33)$$

$$E_1 = \frac{K_1 \cdot a_{C-C}}{A_{C-C}} = \frac{k_{x'} \cdot a_{C-C}}{A_{C-C}} \qquad (34)$$

where K_1 is the elastic stiffness factor of the fiber at the 1 direction, a_{C-C} and A_{C-C} are the C-C bond length and the cross-sectional area of the equivalent fiber, E_1 is the elastic modulus of the fiber at the 1 direction.

The 2 axis for the fiber is defined as the vertical direction of the C-C bond in the σ-plane. The fiber interactions at the 2 direction are mainly reflected through the C-C-C angle bending potential (Fig. 10).

Set K_2 to be the elastic stiffness factor of the fiber at the 2 direction, we obtain,

$$2\cdot\left[2\cdot\frac{1}{2}k_\theta\cdot\delta_\theta^2+\frac{1}{2}k_\theta\cdot(2\delta_\theta)^2\right]=\frac{1}{2}K_2\cdot\delta_2^2=\frac{1}{2}K_2\cdot(2\cdot a_{C-C}\cdot\delta_\theta\cdot\sin\frac{\pi}{6})^2 \quad (35)$$

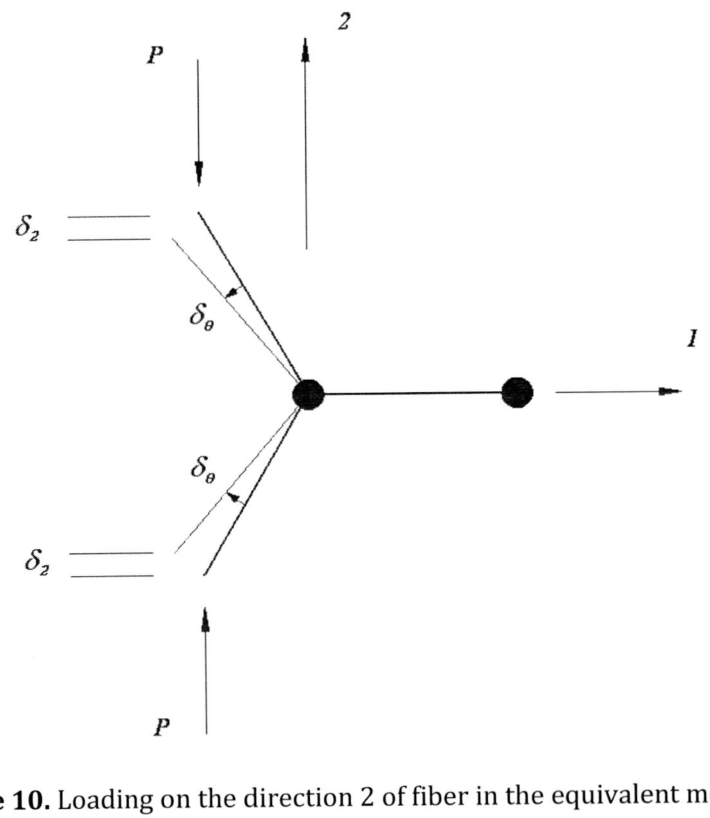

Figure 10. Loading on the direction 2 of fiber in the equivalent model of graphite

$$K_2 = \frac{12 \cdot k_\theta}{a_{C-C}^2} \tag{36}$$

$$E_2 = \frac{1}{2} \frac{K_2 \cdot \sqrt{3} \cdot a_{C-C}}{t \cdot (a_{C-C} + a_{C-C} \cdot \sin\frac{\pi}{6})} = \frac{\sqrt{3}K_2}{3t} = \frac{4\sqrt{3}k_\theta}{3t \cdot a_{C-C}^2} \tag{37}$$

where δ_θ and δ_2 are the C-C-C angle change and the displacement at the 2 direction when the model is loaded at the 2 direction, t is the effective thickness of the equivalent fiber layer, E_2 is the elastic modulus of the fiber at the 2 direction.

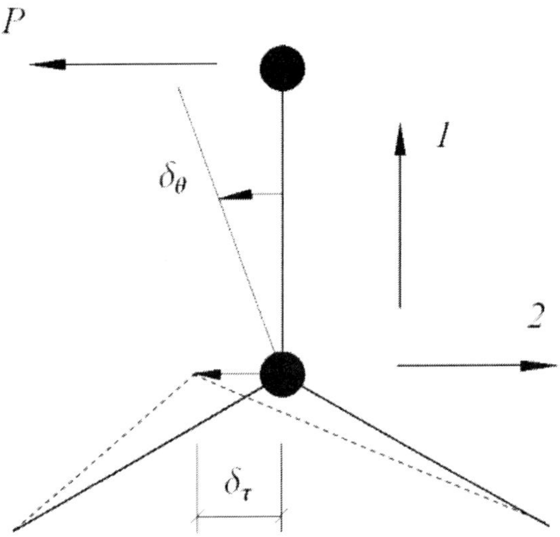

Figure 11. Shearing deformation of the fiber in the equivalent model of graphite

When the fiber layer in the equivalent model is subjected to a shearing force P (Fig. 11), there will be a horizontal displacement $\delta\tau$ for the equivalent fiber element and an angle deflection $\delta\theta$. As to the displacement of the bottom atom in of the figure, it can be obtained,

$$P = 2 \cdot [K_1 \cdot (\delta_\tau \cdot \cos\frac{\pi}{6})] = \sqrt{3}K_1\delta_\tau \tag{38}$$

The strain energy induced by the angle deflection

δθ of a fiber is,

$$2 \cdot (2 \cdot \frac{1}{2}k_{\theta'} \cdot \delta_\theta^2) = \frac{1}{2}P \cdot a_{C-C} \cdot \delta_\theta \tag{39}$$

Substitute equation (38) into (39), we obtain,

$$a_{C-C} \cdot \delta_\theta = \frac{\sqrt{3}K_1\delta_\tau \cdot a_{C-C}^2}{4k_{\theta'}} = \frac{\sqrt{3}K_1}{K_2 / 3}\delta_\tau \tag{40}$$

Therefore, the shearing deformation of the fiber element in the equivalent model is,

$$\bar{\delta}_\tau = \delta_\tau + a_{C-C} \cdot \delta_\theta = (1 + \frac{3\sqrt{3}K_1}{K_2})\delta_\tau \tag{41}$$

According to the definition of the elastic shear modulus:

$$G_{12} = \frac{\tau_{12}}{\gamma_{12}} \tag{42}$$

for the fiber element in the equivalent model, there is:.

$$G_{12} = \frac{P / (t \cdot 2 \cdot a_{C-C} \cdot \cos\frac{\pi}{6})}{\frac{1}{2}\bar{\delta}_\tau / a_{C-C}} = \frac{2P}{\sqrt{3}\bar{\delta}_\tau \cdot t} \tag{43}$$

We substitute equation (38) and (41) into (43), and get

$$G_{12} = \frac{2K_1\delta_\tau}{t\bar{\delta}_\tau} = \frac{2K_1}{t(1 + \frac{3\sqrt{3}K_1}{K_2})} \tag{44}$$

4.4. Test For The Mechanical Constants Of The Monolayer In The Equivalent Model

From equation (34), (37) and (44), we obtain the mechanical constants of the monolayer in the equivalent model

E_1, E_2 and G_{12}, of which E_1 and E_2 are independent quantities, G_{12} is a function of E_1 and E_2 Both E_1 and E_2 are related to MD parameters of the C-C bond energyThere are several empirical potentials and relevant parameters for C-C bond energy. In this section, the following potential energy function and parameters are used to verify the equivalent model provided in this chapter. The Morse potential is employed for the bond stretching action and harmonic potential for the angle bending. The short-range potential caused by the deformation of C-C bond is described as bellow,

$$U_r = K_r(1 - e^{-\beta(r_{ij} - r_0)})^2 \tag{45}$$

$$U_\theta = \frac{1}{2} K_\theta (\theta_{ijk} - \theta_0)^2 \tag{46}$$

where Ur and Uθ are the potentials of bond stretching and angle bending, Kr and Kr are the corresponding force constants. Rij represents the distance between any couple of bonded atoms, θijk represents all the possible angles of bending, r0 and θ0 are the corresponding reference geometry parameters of grapheme.β defines the steepness of the Morse well. The values of all these parameters are listed in Table 1.

Table 1. Parameters for C-C bond in MD

Bond	$K_r = 478.9\text{KJ/mol}$, $\beta = 21.867\text{nm}^{-1}$, $r_0 = 0.142\text{nm}$
Angle	$K_\theta = 418.4\text{KJ/mol}$, $\theta_0 = 120.00°$

Comparing equation (32) and (45), using the data in Table 1, we can obtain:

$$K_1 = k_{x'} = \beta^2 \cdot 2K_r = 760.4 \;\; \mathrm{nN} / \mathrm{nm} \tag{47}$$

We substitute data in Table 1 into equation (36), and get

$$K_2 = \frac{12 \cdot k_{\theta'}}{a_{C-C}^2} = 413.48 \;\; \mathrm{nN} / \mathrm{nm} \tag{48}$$

There are other scholars working on the C-C stretching force constants through experiments or theoretical calculations, who reported the values along the C-C bond like 729nN/nm, 880nN/nm, 708 nN/nm and so on. They also obtained the constants at the direction perpendicular to the C-C bond, 432nN/nm and 398nN/nm (Yang & Zeng, 2006). Noting equation (47) and (48), we can see that the data obtained here based on the current equivalent model are in good agreement with the results from other researchers.

5. ELASTIC PROPERTIES OF GRAPHITE SHEET

5.1. Flexibility Of Monolayer In The Equivalent Model

Graphite sheet can be considered as the network structure formed by the three groups of parallel fibers which are into 60 degree angles with each other. Based on the result of the last section, we can get the axial flexibility of the fiber monolayer,

$$[S]^{(1)} = \begin{bmatrix} \frac{1}{E_1} & -\frac{\nu_{12}}{E_2} & 0 \\ -\frac{\nu_{21}}{E_1} & \frac{1}{E_2} & 0 \\ 0 & 0 & \frac{1}{G_{12}} \end{bmatrix} \tag{49}$$

where it has $\frac{\nu_{12}}{E_2} = \frac{\nu_{21}}{E_1}$. E_1, E_2 and G_{12} can be calculated according to the equation (34), (37) and (44). If we consider ν_{21} to be 0.3 with reference to the parameters of general materials, assume the fiber thickness of 0.34nm, and apply the data in Table 1, the follows can be calculated,

$$E_1 = 1.190 \text{ TPa}, \quad E_2 = 0.702 \text{ TPa}, \quad G_{12} = 0.424 \text{ TPa}$$
$$\frac{\nu_{12}}{E_2} = \frac{\nu_{21}}{E_1} = 0.252 \tag{50}$$

Substituting the values in equation (50) into equation (49), we get the axis flexibility of the monolayer,

$$[S]^{(1)} = \begin{bmatrix} 0.8403 & -0.252 & 0 \\ -0.252 & 1.4245 & 0 \\ 0 & 0 & 2.360 \end{bmatrix} \tag{51}$$

where the unit of the values is 10−3nm2/nN.

Substituting θ=π/3 and the values in equation (51) into equation (28), we get the off-axis flexibility of the 60° and −60° monolayers,

$$\left[\bar{S}\right]^{(60^\circ)} = \begin{bmatrix} 1.2038 & -0.1743 & -0.3444 \\ -0.1743 & 0.9097 & -0.1650 \\ -0.3444 & -0.1650 & 2.6693 \end{bmatrix} \tag{52}$$

$$\left[\bar{S}\right]^{(-60^\circ)} = \begin{bmatrix} 1.2038 & -0.1743 & 0.3444 \\ -0.1743 & 0.9097 & 0.1650 \\ 0.3444 & 0.1650 & 2.6693 \end{bmatrix} \tag{53}$$

The unit of the values is also 10^{-3}nm^2/nN in the last two equations.

5.2. Elastic Properties Of The Equivalent Model Of Graphite

The graphite sheet is a whole layer structure with no delamination and the strain did not change in the thickness. Thus we can use formula (29) to calculate the stiffness,

$$\left[\bar{Q}\right]=\frac{1}{3}\sum_{k=1}^{3}\left[\bar{Q}\right]^{(k)} \tag{54}$$

where $\left[\bar{Q}\right]$ is the converted stiffness of the graphite sheet, $\left[\bar{Q}\right]^{(k)}$ is the stiffness of the kth layer in the equivalent model. The flexibility of the graphite sheet is,

$$\left[\bar{S}\right]=\left[\bar{Q}\right]^{-1}=3/\sum_{k=1}^{3}\left[\bar{Q}\right]^{(k)} \tag{55}$$

Substituting the values in equation (51) ~ (53) into (55), we can obtain,

$$\left[\bar{S}\right]=\frac{3}{\{[S]^{(1)}\}^{-1}+\{\left[\bar{S}\right]^{(60^\circ)}\}^{-1}+\{\left[\bar{S}\right]^{(-60^\circ)}\}^{-1}}=\begin{Bmatrix}1.0251 & -0.2063 & 0\\ -0.2063 & 1.0251 & 0\\ 0 & 0 & 2.4628\end{Bmatrix} \tag{56}$$

where the unit of the values is $10^{-3}nm^2/nN$.

A series of well acknowledged experiment values (Yang & Zeng, 2006) of elastic constants of perfect graphite are listed in Table 2. Data obtained in current work are in good agreement with those results, which justifies the present equivalent model. It should be noticed that there is an obvious error of the S12. However, S12 has little effect on the mechanical properties of the graphite sheet, the error on it will not influence the reasonable application of the current equivalent model in the mechanics analysis of graphite.

Both the result in equation (56) and the data listed in Table 2 indicate that the graphite is in-plane isotropic at plane stress state. And according to the classical theory of composite mechanics, it is known that the π/m laminated plates with m≥3 are in-plane isotropic. We can see that the graphite sheets being in-plane isotropic is mainly due to their special structure of C-C-C angles.

Table 2. Data obtained in current work and elastic constants of perfect graphite crystals from other experiment

Items	S_{11} / 10^{-3} nm^2/nN	S_{12} / 10^{-3} nm^2/nN	S_{66} / 10^{-3} nm^2/nN
Experiment	0.98	-0.16	2.28
Current	1.0251	-0.2063	2.4628
Error	4.60%	28.94%	8.02%

According to the points made above, we try to explain why the elastic properties of CNTs are anisotropic to some extent. One of the possible reasons is that the curling at different curvature from graphene to CNTs makes the C-C-C angles change (e.g. the angles in an armchair CNT with 1nm diameter are not fixed 120°, but about 118 °), for which the quasi-isotropy of graphite sheet is disrupted and the orthotropy introduced. Having an general realization of the CNTs, one should find that the change of the C-C-C angle is obviously related to the change of CNT radius, especially when it is a small tube in diameter. Thus, the orthotropy of CNTs is diameter related, which is in good agreement with the other research results referring to the mechanical properties of CNTs changing due to their various radial size. The changes of C-C-C angles are also somehow related to the chirality of CNTs. However, the extent of the changes of chiral angles is about 0º ~30º, and the change of C-C-C induced by the difference of chiral angles is little when to curl the graphite sheet at the same curvature. It agrees with that many studies reporting that the difference of elastic properties among various chiral CNTs decreases with the increase in the diameter of CNTs.

5.3. Elastic Properties Of The Equivalent Model With Various C-C-C Bond Angles

The C-C-C bond angles will change from the constant 120º to smaller values when the graphite sheet curling to CNTs. In the same way, the angles between the fibers in the equivalent model will change to less than π/3 and the effect of the change on the elastic properties of the model will be investigated in this section. Substituting equation (51) and the angle θ' between the fibers (responding to the changed C-C-C angles) into equation (28), we can obtain the off-axis flexibility of θ' and -θ' monolayers, $[\bar{S}]^{(\theta')}$ and $[\bar{S}]^{(\theta')}$. Substitute those two into equation (55) to get the approximate converted flexibility of the equivalent model (responding to the changed C-C-C angles):

$$[\bar{S}_{\theta'}] = \frac{3}{\{[S]^{(1)}\}^{-1} + \{[\bar{S}]^{(\theta')}\}^{-1} + \{[\bar{S}]^{(-\theta')}\}^{-1}} \tag{57}$$

The elastic modulus of the equivalent model responding to the graphite with changed C-C-C angles, at the direction of 0° fiber or at the vertical direction, can be obtained from equation (57). The variation of the modulus with the changes of the C-C-C angles are displayed in Fig. 12 and Fig. 13, and also displayed the variation of G_{12}, ν_{12} and ν_{21} in Fig.14~Fig.16.

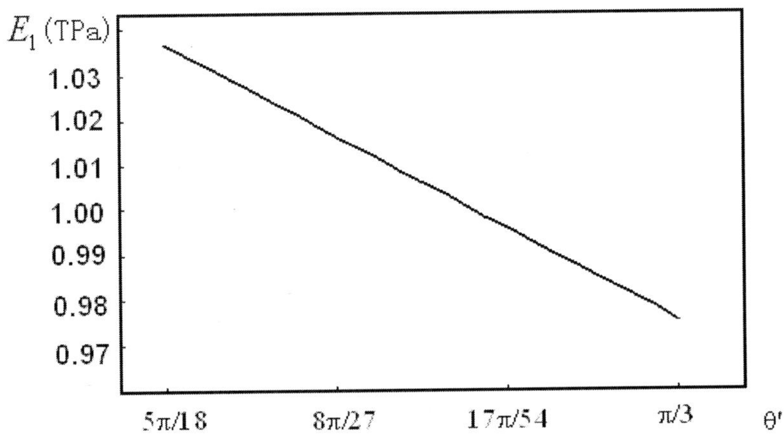

Figure 12. Elastic modulus of the equivalent model of graphene along the 0º fiber

We can see from Fig. 12 and Fig. 13 that the elastic modulus along the 0° fiber of the equivalent model of graphene almost increases linearly, and the elastic modulus at the vertical direction decreases, with the decrease in the C-C-C angle. And Fig. 14 ~ Fig. 16 show that the G_{12}, ν_{12}

and ν_{21} of the equivalent model also changes a lot with the changes of the C-C-C angle.

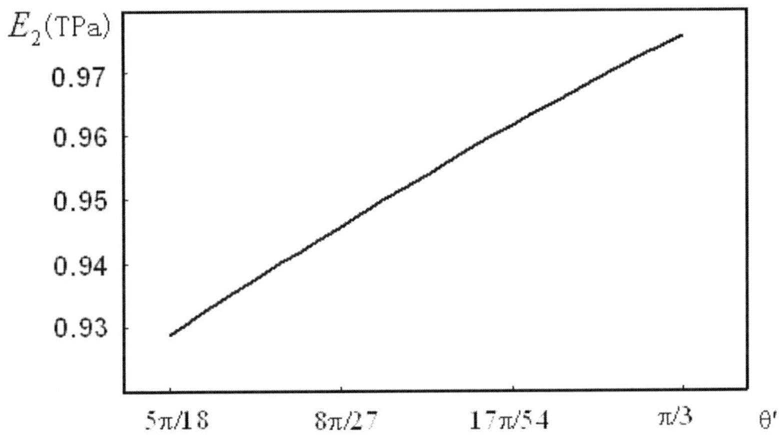

Figure 13. Elastic modulus of the equivalent model of graphene at perpendicular direction to the 0º fiber

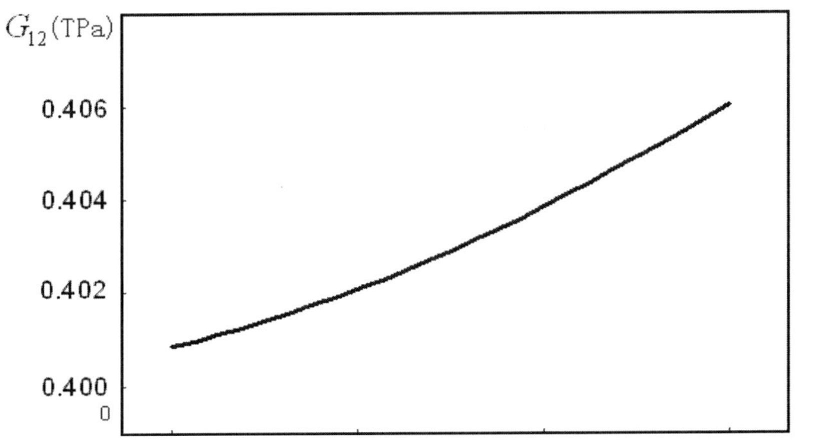

Figure 14. G_{12} of the equivalent model of grapheme

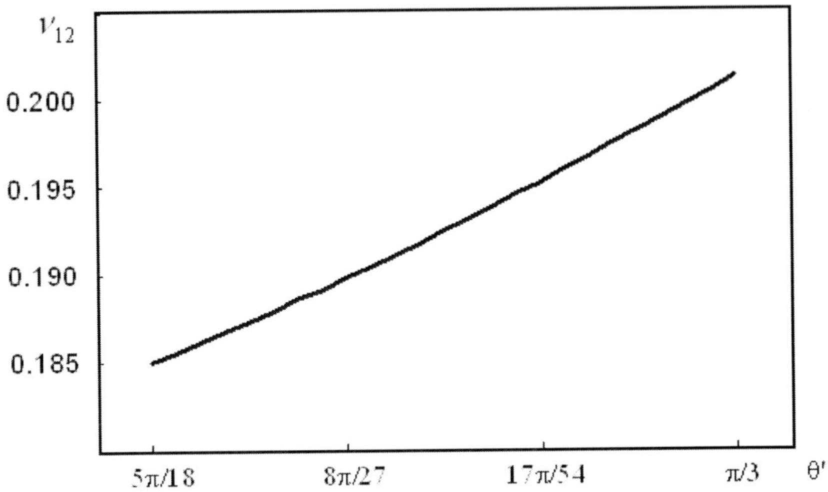

Figure 15. ν_{12} of the equivalent model of grapheme

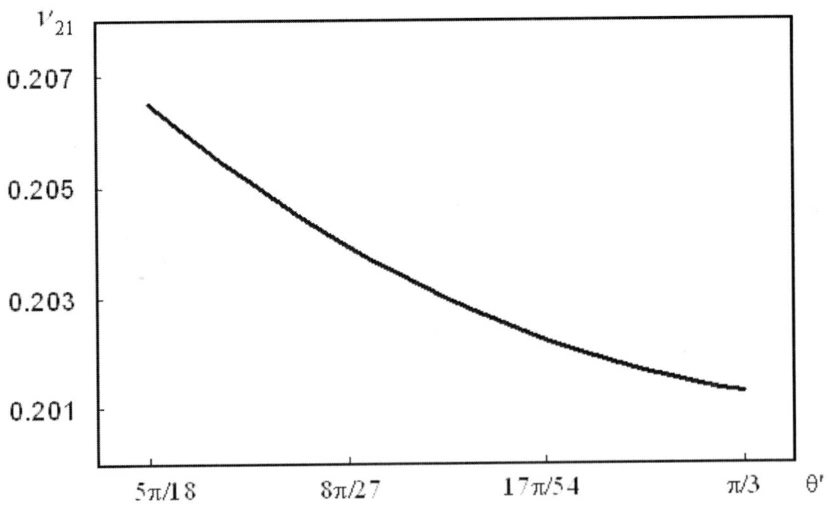

Figure 16. ν_{21} of the equivalent model of grapheme

6. SCALE EFFECT OF ELASTIC PROPERTIES OF CNTS

6.1. Equivalent Model Of Single-Walled Carbon Nanotubes (Swcnts)

CNTs can be considered as curling graphite sheet. The C-C-C bond angles will change from the constant 120º to smaller values when the graphite sheet curling to CNTs and the change of the C-C-C angle is related to the radius of the formed CNTs. According to that, we consider CNTs same as graphene with changed C-C-C angles and the elastic properties of CNTs are consistent with those of graphene with changed C-C-C angles.

6.1.1. Zigzag Swcnts

Taking the zigzag SWCNTs as an example, we study the effect of the changing in diameter of SWCNTs on the value of C-C-C angle. The SWCNT structure and the geometric diagram are shown inFig. 17, with which we can obtain,

$$\sin\angle DBF = \frac{DF}{DB}$$
$$\sin\angle CEF = \frac{CF}{CE} \tag{58}$$

With *DF=CF*, the last equation becomes,

$$\sin\angle DBF = \frac{CE}{DB}\cdot\sin\angle CEF \approx \frac{\sqrt{3}}{2}\sin\angle CEF \tag{59}$$

where the $\angle BDE$ does not change a lot when the graphene curls to the SWCNT so that we make $\frac{CE}{DB}=\frac{DE}{DB}\approx\frac{\sqrt{3}}{2}$.

For an (m, 0) zigzag SWCNT, it has,

$$\angle CEF = \frac{(m-1)\pi}{2m} \tag{60}$$

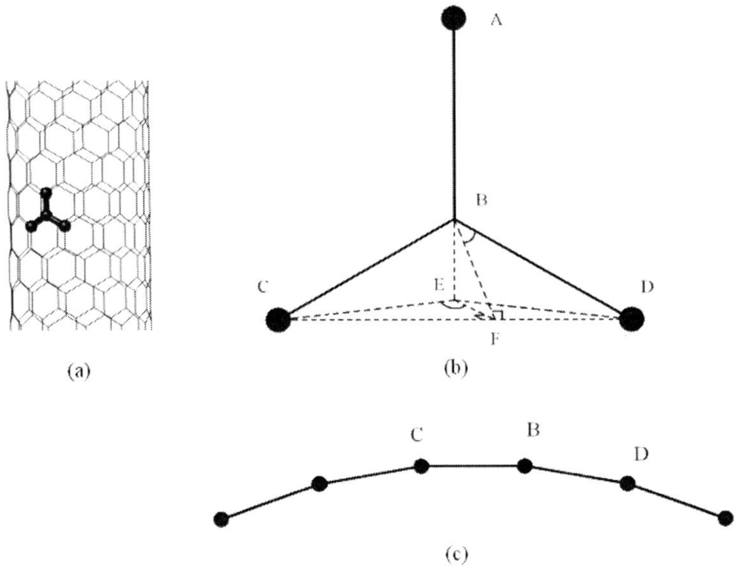

Figure 17. C-C-C angles in the zigzag SWCNTs

The diameter of the tube is,

$$d_{cnt} = 0.0783 \cdot m \quad \text{nm} \tag{61}$$

Substituting equation (61) and (60) into (59), we get,

$$\sin\angle DBF = \frac{\sqrt{3}}{2}\sin\left[\frac{\pi}{2}\left(1-\frac{0.0783}{d_{cnt}}\right)\right] \tag{62}$$

Thus, with the diameter of the zigzag SWCNT provided, setting the AB direction in Fig. 17 as the direction of the 0° fiber, we can calculate the C-C-C angles in zigzag SWCNTs as,

$$\theta' = \text{Arcsin}\left\{\frac{\sqrt{3}}{2}\cdot\sin\left[\frac{\pi}{2}\left(1-\frac{0.0783}{d_{cnt}}\right)\right]\right\} \tag{63}$$

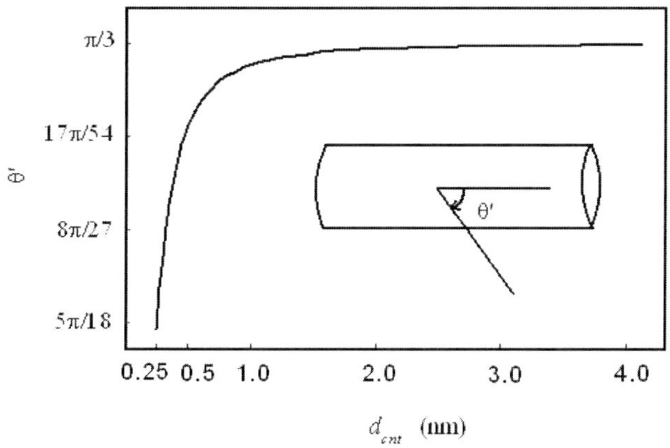

Figure 18. Different C-C-C angles in zigzag SWCNTs with different diameters

The curve drawn from equation (63) is displayed in Fig. 18, which shows that the smaller the tube radius is the more sensitive the changing in C-C-C angle is. For the SWCNTs with diameter 0.4nm, 1.0nm, 2.0nm and 4.0nm, the C-C-C angles in the tubes decrease 7.3%, 1.2%, 0.3% and 0.08% from 120º in the graphene.

6.1.2 Armchair Swcnts

Taking the armchair SWCNTs as another example, we study the effect of the changing in diameter of SWCNTs on the value of C-C-C angle. The SWCNT structure and the geometric diagram are shown inFig. 19, with which we can obtain,

$$\tan\angle DBG = \frac{GD}{GB}$$
$$\tan\angle EBH = \frac{HE}{HB} \tag{64}$$

With *HE=GD*, the last equation becomes,

$$\tan\angle EBH = \frac{HE}{HB} = \frac{GB}{HB}\cdot\tan\angle DBG \approx \frac{\sqrt{3}}{\cos\angle HBG} \tag{65}$$

where the $\angle CBD$ does not change a lot when the graphene curls to the armchair SWCNT so that we make tan $\angle DBG = \tan\frac{\angle CBD}{2} \approx \sqrt{3}$.

For an (m, m) armchair SWCNT, it has,

$$\angle HBG = \frac{\pi}{2m} \tag{66}$$

The diameter of the tube is,

$$d_{cnt} = 0.0783\cdot\sqrt{3}m \quad \text{nm} \tag{67}$$

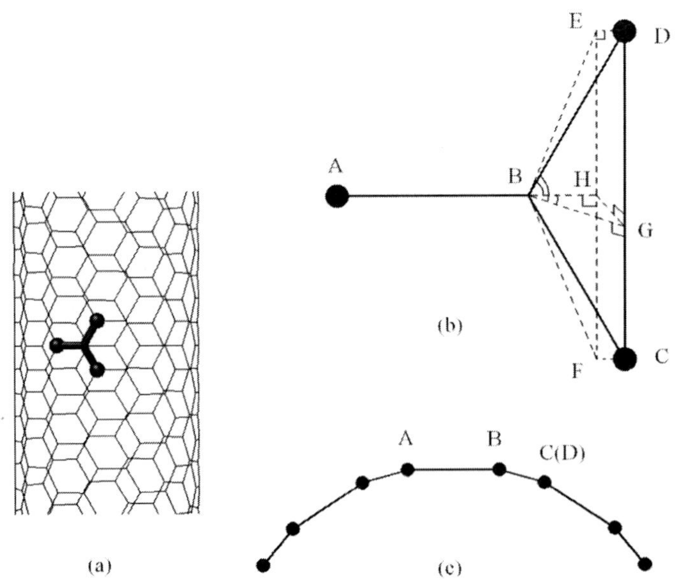

Figure 19. C-C-C angles in the armchair SWCNTs

Substituting equation (67) and (66) into (65), we get,

$$\tan \angle EBH = \frac{\sqrt{3}}{\cos\left(\frac{0.0783\sqrt{3}\cdot\pi}{2d_{cnt}}\right)} \tag{68}$$

Thus, with the diameter of the armchair SWCNT provided, setting the AB direction in Fig. 19 as the direction of the 0º fiber, we can calculate the C-C-C angles in armchair SWCNTs as,

$$\theta' = \text{Arc}\tan\frac{\sqrt{3}}{\cos\left(\frac{0.0783\sqrt{3}\cdot\pi}{2d_{cnt}}\right)} \tag{69}$$

The curve drawn from equation (69) is displayed in Fig. 20, which shows, as for the zigzag SWCNTs, that the smaller the tube radius is the more sensitive the changing in C-C-C angle is. For the armchair SWCNTs with diameter 0.4nm, 1.0nm, 2.0nm and 4.0nm, the C-C-C angles in the tubes decrease 5.9%, 0.94%, 0.23% and 0.06% from 120º in the graphene. Comparing the armchair SWCNTs with the zigzag ones with same diameter, the changing in C-C-C angle in the armchair SWCNTs is smaller.

6.2. Scale Effect Of Elastic Properties Of Swcnts

With the usual continuum model of nanotubes, many researchers use the same isotropic material constants for CNTs with different radius so that the changes of the material properties due to changes in the radial size of CNTs can not be taken into account. In this section, axial deformation of the SWCNTs with different diameter is analyzed with the finite element method to study the effect of the changes of the material properties for CNTs with different radius on the mechanical properties.

According to the theoretical model described above in this chapter, changes in the elastic properties of CNTs are mainly attributed to the changes of the C-C-C angles when to make CNTs by curling the graphene. The C-C-C angles in CNTs are related to the tube radius, while the corresponding C-C-C angles in the graphene induce varying degrees of anisotropy. Thus, we should choose different anisotropic material parameters for different radial size CNTs based on the parameters of the graphene with different C-C-C angles. Some elastic constants of graphite sheet and SWCNTs are listed in Table 3, the values for SWCNTs calculated with equation (57), (63) and (69). The subscript 1 and 2 in the table identify the along-axis direction of SWCNTs and the circumferential direction.

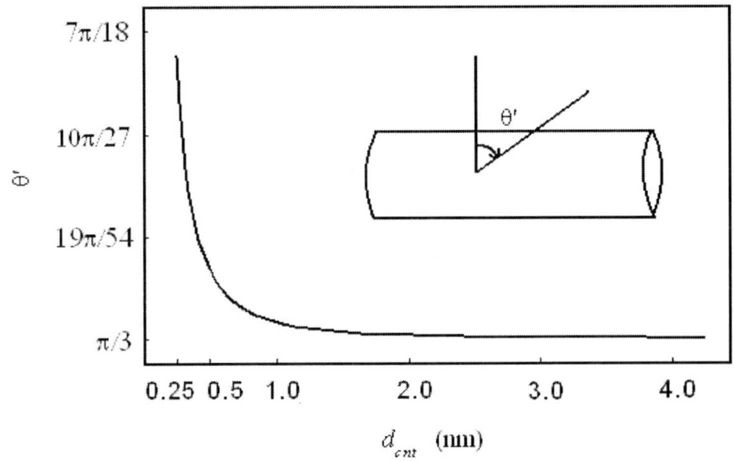

Figure 20. Different C-C-C angles in armchair SWCNTs with different diameters

In the finite element simulation, the same axial strain is applied to each SWCNT, with one end of the tube fixed and at the other end imposed the axial deformation. The length of the tube is 6nm and the axial compression strain is 5%. Two series of the elasticity parameters, the isotropic ones (fromequation (56)) and the anisotropic ones (from equation (57)), are both tried to obtain the axial forces in SWCNTs for comparison. With the results of the FEA, we use the following equation to define the scale effect of SWCNTs,

$$\xi = \frac{R_{\mathrm{ANISO}} - R_{\mathrm{ISO}}}{R_{\mathrm{ISO}}} \tag{70}$$

where R *ANISO* and R *ISO* are the axial forces obtained with the anisotropic parameters and the isotropic ones. The scale effect of SWCNTs with different radial size is shown in Fig. 21.

FEA results show that the anisotropy of the SWCNTs gradually increased with the decreases in the diameter, leading to more and more obvious scale effect. It can be seen from Fig. 21 that compared with the armchair SWCNTs the zigzag ones show more apparent scale effect and the scale effect of SWCNTs with diameter greater than 2nm is negligible (<0.05%), for tubes with any chirality. However, for zigzag SWCNTs with very small diameter, the scale effect could be very obvious up to 4.4%. That is in good agreement with the other researcher's results. Thus, the scale effect should be considered in the accurate calculation about the mechanical behaviour of small SWCNTs.

With the data in Table 3 and the results from Fig. 21 we can see that, for SWCNTs with diameter greater than 2nm, the change of C-C-C angle due to the change in tube diameter is negligible so that the tubes could be treated as isotropic materials with the elastic parameters of graphene. For SWCNTs with diameter smaller than 1nm, the change of C-C-C angle due to the change in tube diameter should not be neglected and it is better for the tubes to be treated as anisotropic materials with the elastic parameters calculated from equation (57), (63) and (69).

7. CONCLUSION

A new equivalent continuum model is presented to theoretically investigate the elastic properties of the graphite sheet. By comparison, the equivalent model can properly reflect the actual elasticity status of graphite sheet. Further more this equivalent model is employed to study the radial scale effect of SWCNTs.

The C-C-C bond angles will change from the constant 120° to smaller values when the graphite sheet curling to CNTs. The change of the C-C-C angle is obviously related to the change in CNT radius. Then the relationship between the anisotropy and the changes of the C-C-C angles of CNTs is deduced. The present theory not only clarify some puzzlement in the basic mechanical research of CNTs, but also lay the foundations for the application of continuum mechanics in the theoretical analysis of CNTs.

Based on above theory the scale effect of CNTs is studied. It is showed that the scale effect of the zigzag CNTs is more significant than the armchair ones. For SWCNTs with diameter greater than 2nm, the change of C-C-C angle due to the change in tube diameter is negligible so that the tubes could be treated as isotropic materials with the elastic parameters of graphene, and the scale effect could also be neglected no mater what chirality they are. However, for SWCNTs with diameter smaller than 1nm, the change of C-C-C angle due to the change in diameter should not be neglected (the scale effect neither) and it is better for the tubes to be treated as anisotropic materials with the elastic parameters calculated from corresponding equations.

Table 3. Elastic constants of graphite sheet and SWCNTs

C-C-C angles			Mechanical parameters					conclusion
Graphene			$E = 0.9755$ TPa , $G = 0.406$ GPa , $\upsilon = 0.201$					Isotropic
SWCNTs			E_1 (TPa)	E_2 (TPa)	G_{12} (TPa)	υ_{12}	υ_{21}	–
	Diameter (nm)	C-C-C angle (°)						
Zig-zag	0.4	55.6298	1.00221	0.956849	0.403246	0.193494	0.202667	Anisotropic
	0.6	57.9833	0.987757	0.967253	0.404666	0.197574	0.201762	Anisotropic
	1.0	59.2586	0.980001	0.972556	0.405521	0.199891	0.201421	Anisotropic
	2.0	59.8129	0.976654	0.974786	0.405908	0.200919	0.201304	Isotropic
	3.0	59.9167	0.976029	0.975199	0.405982	0.201112	0.201284	Isotropic
	4.0	59.9531	0.97581	0.975343	0.406008	0.20118	0.201277	Isotropic
Arm-chair	0.4	63.5548	0.988636	0.954613	0.40873	0.200935	0.208097	Anisotropic
	0.6	61.5715	0.981558	0.966161	0.407195	0.201043	0.204247	Anisotropic
	1.0	60.5640	0.977735	0.972147	0.406448	0.201172	0.202328	Anisotropic
	2.0	60.1408	0.976084	0.974682	0.406142	0.201242	0.201532	Isotropic
	3.0	60.0626	0.975776	0.975152	0.406086	0.201256	0.201385	Isotropic
	4.0	60.0352	0.975667	0.975317	0.406066	0.201261	0.201334	Isotropic

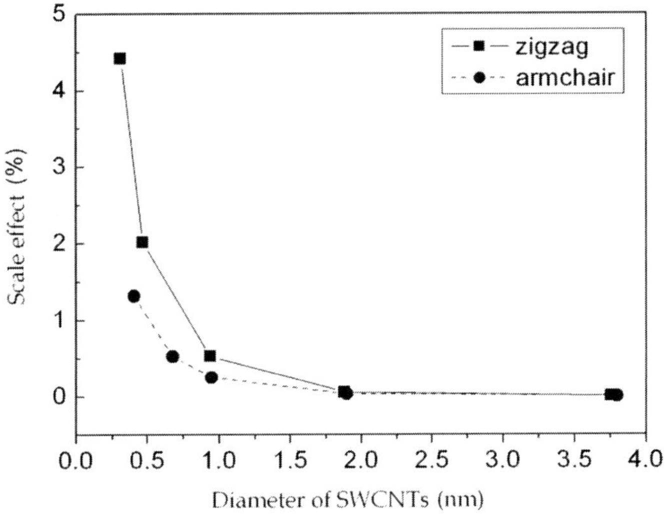

Figure 21. Scale effect of SWCNTs with different diameters

It is theoretically demonstrated that the graphite sheet is in-plane isotropic under plane stress, which is mainly due to their special structure of C-C-C angles. Any deformation of the graphite molecule making changes in the C-C-C angles, e.g. curling, will introduce anisotropic elastic properties. That provides a direction for applying the composite mechanics to the research in the mechanical properties of CNTs, and also has laid an important foundation.

ACKNOWLEDGEMENTS

The authors wish to acknowledge the supports from the Natural Science Foundation of Guangdong Province.

REFERENCES

1. D. W. Brenner, 1990 Empirical potential for hydrocarbons for use in simulating the chemical vapor deposition of diamond films. Phys.Rev. B, 42 15 (November 1990), 9458 9471 , 1098-0121

2. Gao, X. L. , Li, K. , 2003 Finite deformation continuum model for single-walled carbon nanotubes. International Journal of Solids and Structures, 40 26 (December 2003), 7329 7337 , 0020-7683

3. Govindjee, S. , Sackman, J. L. , 1999 On the use of continuum mechanics to estimate the properties of nanotubes. Solid State Commun. 110 4 (March 1999), 227 230 , 0038-1098

4. Harik, V. M. , 2001a Ranges of applicability for the continuum-beam model in the mechanics of carbon-nanotubes and nanorods. Solid State Commun., 120 7-8 , (October 2001), 331 335 , 0038-1098

5. Hu, N. Nunoya, K. Pan, D. Okabe, T. , Fukunaga, H. , 2007 Prediction of buckling characteristics of carbon nanotubes. International Journal of Solids and Structures, 44 20 (October 2007), 6535 6550 , 0020-7683

6. Jin, Y. , Yuan, F. G. , 2003 Simulation of elastic properties of single-walled carbon nanotubes. Compos. Sci. Technol., 63 11 (August 2003), 1507 1515 , 0266-3538

7. Krishnan, A. , Dujardin, E. , Ebbessen, T. W. , Yianilos, P. N. , Treacy, M. M. J. , 1998 Young's modulus of single-walled nanotubes. Phys. Rev. B, 58 20 (November 1998), 14013 14019, 1098-0121

8. Li, C. Y. , Chou, T. W. , 2003 Elastic moduli of multi-walled carbon nanotubes and the effect of van der Waals forces. Compos. Sci. Technol., 63 11 (August 2003), 1517 1524 , 0266-3538

9. Liu, J. Z. , Zheng, Q. S. , Jiang, Q. , 2001 Effect of a Rippling Mode on Resonances of Carbon Nanotubes. Phys. Rev. Lett., 86 21 (May 2001), 4843 4846 , 0031-9007

10. Lu, J. P. , 1997 Elastic properties of carbon nanotubes and nanoropes. Phys. Rev. Lett., 79 7 (August 1997), 1297 1300 , 0031-9007

11. Mayo, S. L. , Olafson, B. D. , I. I. I. Goddard, W. A. , 1990 Dreiding: a generic force field for molecular simulations. J Phys Chem, 94 26 (December 1990), 8897 8909 , 1932-7447

12. Poncharal, P. , Wang, Z. L. , Ugarte, D. , W. A. de Heer, 1999 Electrostatic deflections and electromechanical resonances of

carbon nanotubes. Science, 283 5407 (May 1999), 1513 1516 , 0036-8075

13. Shen, H. S. , 2004 Postbuckling prediction of double-walled carbon nanotubes under hydrostatic pressure. International Journal of Solids and Structures, 41 9-10 , (May 2004), 2643 2657 , 0020-7683

14. Sudak, L. J. , 2003 Column buckling of multiwalled carbon nanotubes using nonlocal continuum mechanics. J. Appl. Phys., 94 11 (November 2003), 7281 7287 , 0021-8979

15. Wang, C. Y. , Ru, C. Q. , Mioduchowski, A. , 2003a Axially compressed buckling of pressured multiwall carbon nanotubes. International Journal of Solids and Structures, 40 5 (July 2003), 3893 3911 , 0020-7683

16. Wang, Q. , Varadan, V. K. , Quek, S. T. , 2006 Small scale effect on elastic buckling of carbon nanotubes with nonlocal continuum models. Physics Letters A, 357 2 (September 2006), 130 135 , 0375-9601

17. Wang, Q. , Varadan, V. K. , 2007 Application of nanlocal ealstic shell theory in wave propagation analysis of carbon nanotubes. Smart Mater. Struct., 6 1 February 2007), 178 190 , ISSN

18. Xin, H. , Han, Q. , Yao, X. H. , 2007 Buckling and axially compressive properties of perfect and defective single-walled carbon nanotubes. Carbon, 45 13 (November 2007), 2486 2495 , 0008-6223

19. Xin, H. , Han, Q. , Yao, X. H. , 2008 Buckling of defective single-walled and double-walled carbon nanotubes under axial compression by molecular dynamics simulation. Compos. Sci. Technol., 68 7-8 , (June 2008), 1809 1814 , 0266-3538

20. Yang, X. G. , Zeng, P. , 2006 Numerical simulation of anisotropic mechanical Properties of nano-graphite crystals. Journal of basic science and engineering, 14 3 Sep tember 2006), 375 383 , 1005-0930

21. Yao, X. H. , Han, Q. , 2007 Postbuckling prediction of double-walled carbon nanotube under axial compression. Eur. J. Mech.A-Solid, 26 1 (January-February 2007), 20 32 , 0997-7538

22. Yao, X. H. , Han, Q. , 2008 Torsional buckling and postbuckling equilibrium path of double-walled carbon nanotubes. Compos. Sci. Technol., 68 1 (January 2008), 113 120 , 0266-3538

23. Yao, X. H. , Han, Q. , Xin, H. , 2008 Bending buckling behaviors of single- and multi-walled carbon nanotubes. Computational Materials Science, 43 4 (October 2008), 579 590 , 0927-0256

24. Zhang, Y. Q. , Liu, G. R. , Xie, X. Y. , 2005 Free transverse vibrations of double-walled carbon nanotubes using a theory of nonlocal elasticity. Phys. Rev. B, 71 (May 2005), 195404 1-7, 1098-0121

25. Yang, X. G. , Zeng, P. , 2006 Numerical simulation of anisotropic mechanical Properties of nano-graphite crystals. Journal of basic science and engineering, 14 3 Sep tember 2006), 375 383 , 1005-0930

2

NICKEL SUPPORTED ON NITROGEN-DOPED CARBON NANOTUBES AS HYDROGEN OXIDATION REACTION CATALYST IN ALKALINE ELECTROLYTE

Zhongbin Zhuang[1,2,3], Stephen A. Giles[2,4], Jie Zheng[2,3], Glen R. Jenness[2,4], Stavros Caratzoulas[2,4], Dionisios G. Vlachos[2,4] & Yushan Yan [2,3]

[1] State Key Lab of Organic-Inorganic Composites, Beijing University of Chemical Technology, Beijing 100029, China.

[2] Department of Chemical and Biomolecular Engineering, University of Delaware, Newark, Delaware 19716, USA.

[3] Center for Catalytic Science and Technology, University of Delaware, Newark, Delaware 19716, USA.

[4] Catalysis Center for Energy Innovation, University of Delaware, Newark, Delaware 19716, USA.

The development of a low-cost, high-performance platinum-group-metal-free hydroxide exchange membrane fuel cell is hindered by the lack of a hydrogen oxidation reaction catalyst at the anode. Here we report that a composite catalyst, nickel nanoparticles supported on nitrogen-doped carbon nanotubes, has hydrogen oxidation activity similar to platinum-group metals in alkaline electrolyte. Although nitrogen-doped carbon nanotubes are a very poor hydrogen oxidation catalyst, as a support, it increases the catalytic performance of nickel nanoparticles by a factor of 33 (mass activity) or 21 (exchange current density) relative to unsupported nickel nanoparticles. Density functional theory calculations indicate that the nitrogen-doped support stabilizes the nanoparticle against reconstruction, while nitrogen located at the edge of the nanoparticle tunes local adsorption sites by affecting the *d*-orbitals of nickel. Owing to its high activity and low cost, our catalyst shows significant potential for use in low-cost, high-performance fuel cells.

The hydrogen economy provides an efficient and environmentally friendly pathway to store and consume energy[1, 2]. Fuel cells, especially the recently developed polymer electrolyte membrane fuel cells, are considered as the most promising device to convert the chemical energy of hydrogen to electricity[3]. The hydrogen fuel cells are based on two half-cell reactions: hydrogen oxidation reaction (HOR) at the anode and oxygen reduction reaction (ORR) at the cathode. To date, Pt is the most active catalyst for both HOR and ORR[4]. However, the commercialization of fuel cells is hindered by the high price of Pt (~\$50 g^{-1}). Although the total content of platinum-group metals (PGMs) in the state-of-the-art proton exchange membrane fuel cell (PEMFC) stacks has decreased significantly in the past decades, more than 0.137 g_{Pt} kW^{-1} is still needed[5]. One promising approach to reduce the cost of fuel cells is to switch the operating environment from an acidic to a basic one (that is, a hydroxide exchange membrane fuel cell, HEMFC), thus opening up the possibility of using PGM-free catalysts and other cheaper components[6]. For the cathode of the HEMFC, some PGM-free and metal-free ORR catalysts have been developed that show comparable activity to Pt in alkaline media[7, 8, 9, 10]. However, for the anode side, only a few PGMs (for example, Pt, Ir and Pd) show adequate activity[11, 12]. The HOR catalysed by Pt is very fast in acidic conditions so that a very low loading of the Pt catalyst could be used relative to the cathode side in PEMFCs. However, the HOR activities of PGMs are ~100 times slower in alkaline solutions[11, 13]. As a result, a much higher loading of the HOR catalyst is required (0.4 $mg_{Pt}\,cm^{-2}$ in a HEMFC compared with 0.03 $mg_{Pt}\,cm^{-2}$ in a PEMFC) to achieve similar performance[5, 14]. Thus, it is highly desirable to develop PGM-free anode catalysts for the HOR in alkaline electrolyte.

Unlike its reverse reaction (hydrogen evolution reaction, HER)[15, 16, 17], only a few PGM-free HOR catalysts have been reported. One possibility is to use Raney Ni as the HOR catalyst in liquid alkaline fuel cells[18, 19, 20]. However, it is functional only under very high alkalinity (6 M KOH) while the activity remains low. It is not catalytically active for a HEMFC, which can be mimicked as 0.1–1 M KOH[21]. Efforts have been made to improve the HOR activity of the Ni-based catalyst in the last decade. Ni alloys, such as NiMo and NiTi, have been shown to enhance the HOR activity[20]. Our recent work has also shown that electrochemically deposited NiCoMo on an Au substrate has a high HOR activity[22]. Zhuang and co-workers decorated Ni particles with CrO_x to weaken the Ni–O bond and stabilize the Ni catalysts. A HEMFC incorporating this PGM-free catalyst has been fabricated, and it exhibits a peak power density of 50 mW cm^{-2} (ref. 21) Although the power density is still low (compared with the peak power density of more than 1,000 mW cm^{-2} for PEMFCs), it demonstrates the possibility to fabricate low-cost PGM-free fuel cells. However, their activities are still incomparable with PGM-based catalysts.

In the current study, we synthesize Ni nanoparticles supported on N-doped carbon nanotubes (Ni/N-CNT) by a wet chemical method, and the Ni/N-CNT shows a high HOR catalytic activity in 0.1 M KOH solution. N-CNTs are not only the support for the Ni nanoparticles, but also a promoter for the catalytic activity. Owing to its special electronic properties, the N-doped carbon structure has been employed to enhance the catalytic activity of the methanol oxidation reaction and the ORR[23]. We use N-CNT as a support for the HOR catalyst and demonstrate that, relative to Ni nanoparticles, the mass activity and exchange current density of Ni/N-CNT increases by a factor of 33 and 21, respectively. To understand the interaction between the Ni nanoparticle and the N-CNT support, density functional theory (DFT) calculations are undertaken. The DFT calculations indicate that, when nitrogen dopants are present at the edge of the nanoparticle, the Ni nanoparticle is stabilized on the support and locally activated for the HOR because of modulation of the Ni *d*-orbitals. Using the volcano relationship between the HOR activity and the hydrogen adsorption energy, predicted first-principles exchange current densities of the model systems are in good agreement with the measured exchange current densities of the experimental catalysts. Owing to its high activity and low cost, Ni/N-CNT has great potential to be used as the anode in HEMFCs, thereby finally bringing to fruition a high-performance and low-cost PGM-free HEMFC.

1. RESULTS

1.1 Catalyst synthesis and characterization

A two-step approach was developed to synthesize the Ni/N-CNT hybrid catalysts. First, Ni was selectively grown on mildly oxidized multiwalled

CNTs by reducing nickel salt in an aqueous solution. Second, ammonia and hydrazine were added and then subjected to a hydrothermal treatment at 150 °C. This step led to better crystallization of the Ni nanoparticles, partial reduction of the oxidized CNTs and, more importantly, doping of nitrogen into the CNTs.

The scanning electron microscopy (SEM, Fig. 1a) and transmission electron microscopy (TEM,Fig. 1b) images clearly show that the nanoparticles with an average size of 20 nm are selectively grown on the surface of the CNT. The diffraction rings in the selected area electron diffraction (inset of Fig. 1b) pattern can be either indexed to Ni or CNT. The energy-dispersive X-ray spectroscopy (EDS, Supplementary Fig. 1) also indicated the presence of the Ni component. The enlarged TEM image (Fig. 1c) shows that all the nanoparticles are directly supported on the CNT. Moreover, the high-resolution TEM (HRTEM, Fig. 1d) image clearly indicates the layered structure of the CNT wall, with an interplanar spacing of 0.35 nm, which is similar to the (002) plane of graphite. The lattice fringe of the nanoparticle has an interplanar spacing of 0.21 nm and is ascribed to the (111) plane of Ni.

The X-ray diffraction pattern (Fig. 2a) further confirms the mixed Ni and CNT composition of the catalyst. The loading of Ni on carbon was calculated as ~70 wt % by thermogravimetric analysis (TGA, method shown in Supplementary Fig. 2). The N doping was examined with X-ray photoelectron spectroscopy (XPS, Fig. 2b). A peak at *ca*. 400 eV clearly exists in the high-resolution N 1*s* XPS spectrum (inset of Fig. 2b), confirming the N doping. It can be deconvoluted into two peaks, 399.0 and 400.3 eV, corresponding to pyridinic N and pyrrolic N in doped CNT, respectively[24, 25]. The N-doping amount to carbon was calculated from the XPS spectrum as 2.4 at %.

1.2 Electrochemical HOR performance

The electrocatalytic activity for the HOR was investigated by rotating disk electrode measurements using a standard three-electrode system in 0.1 M KOH. The catalyst was uniformly cast on a 5-mm glassy carbon electrode with a Ni loading of 0.25 mg_{Ni} cm^{-2}. Undoped CNT supported Ni nanoparticles (represented as Ni/CNT) and unsupported Ni nanoparticles (represented as Ni) catalysts (corresponding TEM images and X-ray diffraction patterns are shown in Supplementary Figs 3 and 4, respectively) with the same loading were also studied for comparison.

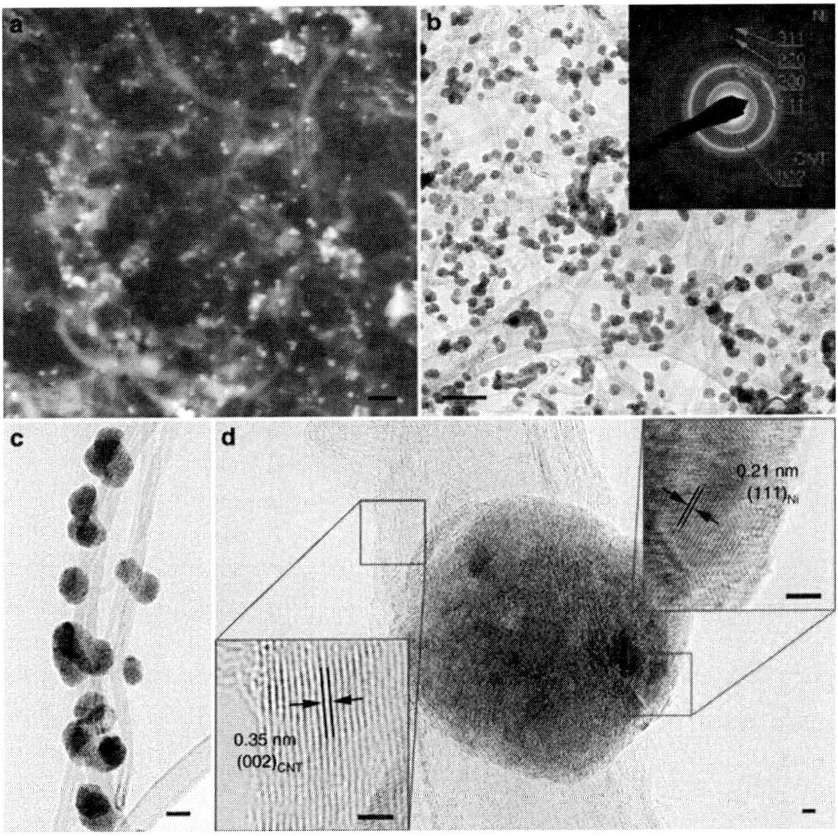

Figure 1. Electron microscopy of the Ni/N-CNT. (**a**) SEM image. Scale bar, 100 nm. (**b**) TEM image. Scale bar, 100 nm. Inset is selected area electron diffraction pattern. (**c**) A magnified TEM image. Scale bar, 20 nm. (**d**) HRTEM images of nickel particle and CNT, respectively. Scale bars, 2 nm.

Figure 3a shows the polarization curve of the catalysts in H_2-saturated electrolyte. The anodic current above 0 V (versus the reversible hydrogen electrode, same hereafter) is assigned to the oxidation of H_2. The catalytic activity for HOR on the three Ni-based catalysts follows the sequence of Ni/N-CNT>Ni/CNT>Ni. The current density of Ni/N-CNT is much higher than either Ni/CNT or Ni and shows an onset potential as low as 0 V. The bare N-CNT shows almost no catalytic activity for HOR, indicating that the enhanced catalytic activity for Ni/N-CNT comes from the synergetic effect of the combined structure. The Ni/N-CNT catalyst was also tested in an Ar-saturated electrolyte (Supplementary Fig. 5) and showed no anodic current, confirming the catalytic reaction of HOR.

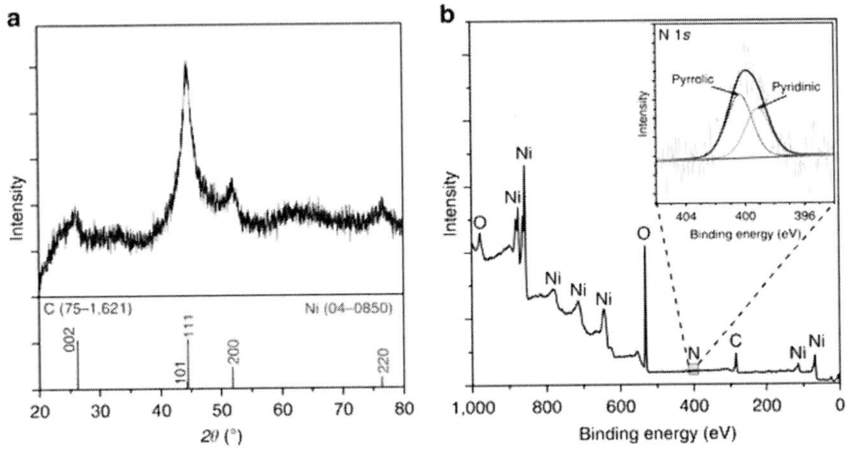

Figure 2: X-ray diffraction pattern and XPS spectra of the Ni/N-CNT. (**a**) X-ray diffraction pattern. The standard pattern of Ni (JCPDS card No. 04-0850) and graphite (JCPDS card No. 75-1621) are shown beneath the plot. (**b**) XPS spectrum. The inset is the high-resolution N 1*s* XPS spectrum.

The polarization curves at different rotating speeds have also been studied (Fig. 3b). The current density increases with increasing rotating speed owing to improved mass transport. The overall current density is under mixed kinetic-diffusion control. The diffusion-limited current density (j_d) for a rotating disk electrode is described by the Levich equation as a function of rotating speed (ω):

$$j_d = Bc_0\omega^{1/2} = 0.62nFD^{2/3}\upsilon^{-1/6}c_0\omega^{1/2} \quad (1)$$

where B is the Levich constant, c_0 is the solubility of H_2 in 0.1 M KOH[26]. The Levich constant is a function of the diffusivity of H_2 (D), number of the electrons (n) transferred in the HOR and kinematic viscosity of the electrolyte (ν). The overall current density (j) can be deconvoluted into kinetic (j_k) and diffusional (j_d) components following the Koutecky–Levich equation:

$$\frac{1}{j}=\frac{1}{j_{\mathrm{k}}}+\frac{1}{j_{\mathrm{d}}}=\frac{1}{j_{\mathrm{k}}}+\frac{1}{Bc_0\omega^{1/2}} \tag{2}$$

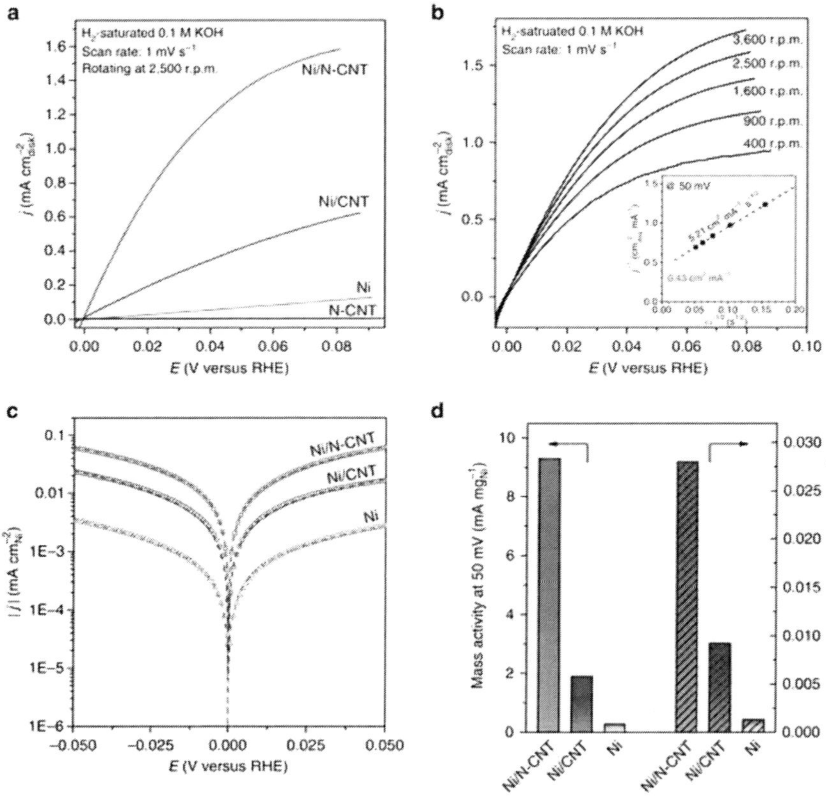

Figure 3. HOR performances. (**a**) Polarization curves of Ni/N-CNT, Ni/CNT, Ni (all of the three catalysts with a loading of 0.25 $mg_{Ni}\,cm^{-2}$) and N-CNT (0.1 $mg_{C}\,cm^{-2}$) catalysts in H_2-saturated 0.1 M KOH at a scan rate of 1 $mV\,s^{-1}$ and rotating speed of 2,500 r.p.m. (**b**) Polarization curves of Ni/N-CNT in H_2-saturated 0.1 M KOH at a scan rate of 1 $mV\,s^{-1}$ and various rotating speeds. Inset is the Koutecky–Levich plot at an overpotential of 50 mV. (**c**) HOR/HER Tafel plots of the specific current density on Ni/N-CNT, Ni/CNT and Ni in H_2-saturated 0.1 M KOH. The dashed lines indicate the Butler–Volmer fitting. (**d**) Mass activity at 50 mV (unpatterned) and exchange current density (patterned) of the Ni/N-CNT, Ni/CNT and Ni, respectively.

A linear relationship between the inverse of j at 50 mV and $\omega^{1/2}$ is observed in the Koutecky–Levich plots (inset of Fig. 3b). The calculated slope is 5.21 $cm^2\,mA^{-1}\,s^{-1/2}$, which is reasonably close to the theoretical number (4.87 $cm^2\,mA^{-1}\,s^{-1/2}$, for the two-electron transfer of HOR)[13]. The intercept of the extrapolated line corresponds to the inverse of the purely kinetic current density, and the j_k for Ni/N-CNT is 2.33 mA cm_{disk}^{-2} at 50 mV overpotential. Normalizing by the metal loading of the catalyst, the mass activity of the Ni/N-CNT is 9.3 mA mg_{Ni}^{-1} at 50 mV overpotential, shown in Fig. 3d and Supplementary Table 1 along with that of Ni/CNT and Ni. The mass activity of our synthesized unsupported Ni nanoparticles (0.28 mA mg_{Ni}^{-1}) is comparable to the activities of Raney Ni reported in the literature[27, 28, 29, 30, 31, 32]. By contrast, the mass activity of the Ni/N-CNT is 33 times as high as the pure Ni nanoparticle, making Ni/N-CNT one of the most active PGM-free catalysts for HOR (Supplementary Table 2 lists the HOR mass activity of the PGM-free catalyst reported in the literature).

To understand the intrinsic activity of the Ni/N-CNT catalyst, we normalized the current by the electrochemical surface area (ECSA). The ECSAs of each Ni-based catalyst were measured in Ar-saturated 0.1 M KOH using a cyclic voltammetry method and were calculated from the OH desorption region of Ni subtracted by HER current using a charge density of 514 μC cm_{Ni}^{-2} for one monolayer of OH adsorption on Ni (Supplementary Fig. 6)[33, 34]. Figure 3c shows the ECSA-normalized kinetic currents as a function of the overpotential. The HOR kinetic current was obtained from the Koutecky–Levich equation (equation 2). The exchange current density (j_0, shown in Fig. 3d and Supplementary Table 1) was obtained by fitting the ECSA-normalized HOR/HER kinetic current density to the Bulter–Volmer equation:

$$j_k = j_0\left(e^{\frac{\alpha F}{RT}\eta} - e^{\frac{(\alpha-1)F}{RT}\eta}\right) \tag{3}$$

Here α is the charge transfer coefficient, F is Faraday's constant, R is the universal gas constant, T is the temperature and η is the overpotential[13]. All the curves can be fitted with a value of α in the range of 0.4~0.5, indicating a good symmetry for the HOR and HER branches. The j_0 values obtained from linear fitting of micropolarization regions (−10–10 mV, see Supplementary Fig. 7 and Supplementary Table 1) are consistent with the values of j_0 obtained from Bulter–Volmer fitting. The Ni/N-CNT has an exchange current density of 0.028 mA cm_{Ni}^{-2}, which is three times higher than Ni/CNT and 21 times higher than Ni. The exchange current density for Ni/N-CNT is among the best PGM-free HOR catalysts and is comparable to some PGM catalysts, such as Pd (Supplementary Table 3)[11, 35]. Although the HOR activity of the Ni/N-CNT is still over one order of magnitude less than the state-of-the-art Pt catalyst, the ultralow price of Ni makes it promising for commercial applications.

1.3 Theoretical investigation

On the basis of a Wulff construction, fcc metals, such as Ni, are known to prefer a cuboctahedral geometry, exposing the {111} and {100} facets[36]. To model the nanoparticle, a 13-atom Ni nanocluster (Ni_{13}) was chosen, as this was the smallest of the cuboctahedral 'magic numbers' (a larger nanoparticle, Ni_{37}, was also studied and is discussed below)[37]. Because of the relatively large size of both the Ni nanoparticles and carbon nanotubes used experimentally, previous studies suggest that the curvature of the CNT support has a minimal effect[38]; as such, the CNT support is approximated with a graphene sheet to reduce the complexity of the model. This results in a Ni_{13} nanoparticle being deposited on both graphene (Ni/graphene) and nitrogen-doped graphene (Ni/N-graphene), with both the Ni_{13} nanoparticle and the support being allowed to relax. We have investigated the effect of the location of the nitrogen dopant relative to the nanoparticle by considering nitrogen at the centre (N_c) and nitrogen at the edge (N_e), as shown inSupplementary Fig. 8.

In alkaline solutions, the HOR follows either the Tafel–Volmer or the Heyrovsky–Volmer mechanism. The possible elementary steps comprising the HOR are shown in Equations 4, 5, 6.

$$\text{Tafel}: H_2 + 2^* \rightarrow 2H_{ad} \quad (4)$$

$$\text{Heyrovsky}: H_2 + OH^- + ^* \rightarrow H_{ad} + H_2O + e^- \quad (5)$$

$$\text{Volmer}: H_{ad} + OH^- \rightarrow H_2O + e^- + ^* \quad (6)$$

The key intermediate is the adsorbed hydrogen (H_{ad}) on the catalyst surface, which leads to the existence of a Sabatier volcano relationship with the hydrogen-binding energy of the catalyst as the descriptor[39].

Although mechanistically the H_{ad} written in the elementary steps above is likely the overpotential deposited hydrogen (H_{OPD}), the HOR/HER activity seems to correlate well to the electrochemical adsorption energy of the underpotential deposited hydrogen (H_{UPD}), which is shown to be similar to the chemical adsorption energy of hydrogen[40, 41, 42]. In this study we calculate chemical adsorption energy of hydrogen on Ni_{13} nanoclusters.

The volcano relationship involving adsorbed hydrogen has been well established in acidic solutions[43]. While some debate remains on the adequacy of using the hydrogen-binding energy as a descriptor in alkaline media, there is significant evidence that the hydrogen-binding energy remains the key descriptor of activity[12, 44, 45]. This allows for the implementation of a hybrid data-driven/first-principles model to evaluate the efficacy of different catalysts and to estimate activity compared with single metals. This hybrid approach was employed successfully in propane total oxidation[46]. Details of the construction of the hybrid activity model are given in the Supplementary Methods.

The heterogeneity of the hydrogen-binding sites in the supported Ni nanocluster model system results in a distribution of binding energies shown in Fig. 4a. Here we see a large amount of variability among the metal-support systems at high binding energies, with an unsupported Ni_{13}nanocluster and Ni/graphene having a large number of sites that adsorb hydrogen too strongly to be active for HOR. On the other hand, Fig. 4a shows that a Ni nanoparticle supported on N_e-graphene does not possess large binding energies as in the unsupported Ni and the Ni/graphene systems.

The differences in the binding energies between the unsupported Ni_{13} nanocluster and the supported Ni_{13} arise from either an electronic (that is, charge transfer) or a geometric (that is, relaxation) origin[47]. For the strong binding sites, the nanocluster undergoes reconstruction as shown in Supplementary Fig. 9. The relative contributions of the electronic and the geometric support effects are decoupled by considering a nanocluster which is treated as rigid during hydrogen binding. For a rigid nanocluster, any differences in the binding energies result from the differences in the electronic structure of the Ni_{13} nanocluster because of the support. We define a relaxation energy, E_{relax}, as

$$E_{relax} = \Delta E_{H}^{rig} - \Delta E_{H}^{rel} \tag{7}$$

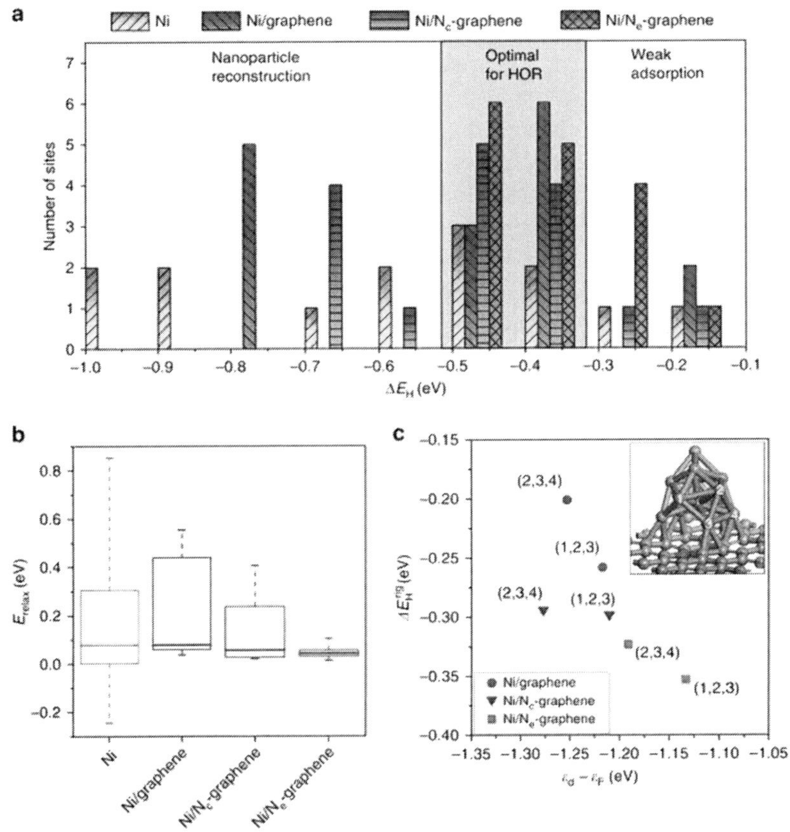

Figure 4. Computational results. (**a**) Distribution of site-dependent hydrogen-binding energies for each model system. (**b**) Distribution of relaxation energies for each model system on hydrogen-binding to each site. The ends of the dashed line represent the minimum and maximum values of E_{relax} for each model structure. These ranges collectively represent the distribution of E_{relax} values for each hydrogen adsorption site on the Ni_{13}cluster. The bottom of the box represents the first quartile (that is, splits the lowest 25% of relaxation energies from the highest 75%) and the top of the box represents the third quartile (that is, splits the highest 25% of relaxation energies from the lowest 75%). Bold horizontal band is the median value. (**c**) Shifts in the *d*-band centre with respect to the Fermi level and binding energy at adjacent Ni sites (1,2,3) and (2,3,4) for Ni/graphene, Ni/N_c-graphene and Ni/N_e-graphene. Inset is graphical depiction of the sites (Ni/N_e-graphene as an example). The (1,2,3) site represents the hollow site in coordination with the #1, #2 and #3 Ni atoms. The (2,3,4) site represents the hollow site in coordination with the #2,#3 and #4 Ni atoms. Blue, grey and red spheres represent Ni, C and N atoms, respectively.

where ΔE_H^{rig} is the hydrogen-binding energy on the rigid nanocluster, and ΔE_H^{rel} is the hydrogen-binding energy on the relaxed nanocluster. Here a larger E_{relax} implies a larger degree of Ni reconstruction. Figure 4b shows E_{relax} for all Ni-binding sites for the systems considered. The unsupported nanocluster is the least stable during hydrogen binding. With the exception of a fraction of binding sites, the nitrogen-doped supports exhibit better stabilization of the Ni nanocluster. In agreement with its lack of strongly binding sites (*cf.* Fig. 4a), Ni/N_e-graphene exhibits local relaxation with minimal reconstruction. In addition, when considering the rigid nanocluster, the difference in the average hydrogen-binding energy of Ni/N_c-graphene and Ni/N_e-graphene is less than 0.01 eV (versus 0.09 eV difference for the relaxed nanocluster). Thus, we observe that the rigid nanocluster's binding characteristics are not greatly affected by the support (N_c-graphene or N_e-graphene).

Next, we examine what fundamental electronic properties of the Ni_{13} nanocluster are being affected by the support. A commonly employed electronic descriptor for the binding energies on metal surfaces is the energy of the *d*-band centre, ε_d, relative to the Fermi level, ε_F (ref. 48). The*d*-band centre describes the relative filling of the hybridized (*d*–*s*) and antibonding (*d*–*s*)* states of the Ni–H system. For an adsorbate, such as atomic hydrogen with a low-lying electronic state, the stabilization of the bonding states between the metal *d*-states and the hydrogen *s*-states decreases when the *d*-band of the metal is shifted up in energy ($\varepsilon_d - \varepsilon_F$ is more positive) because of the energy mismatch between the metal *d*-band and the hydrogen *s*-state. As a result, the resulting antibonding part of the band is pushed up just above the Fermi level, which leads to stronger binding between the metal atoms and the adsorbate. In Fig. 4c, we see an upshift in the*d*-band centre of adjacent Ni-binding sites when an edge nitrogen atom is present. Owing to these sites originally lying in the weak binding region of Fig. 4a, the upshift in the *d*-band centre consequently activates these sites adjacent to N_e for the HOR by strengthening the binding energy by 0.1 eV. XPS valence band spectra, which are related to the *d*-band centres, were also studied (Supplementary Fig. 10)[49]. Thus, while the geometric effect of the support is responsible for the decrease in strong binding sites, edge nitrogen further tunes local binding sites by modulating the electronic properties of Ni.

Using the hybrid activity model for the HOR, we can now predict the intrinsic exchange current density of each of the model systems that were investigated (Supplementary Fig. 11 and Supplementary Note 1). In agreement with experimental measurements, Fig. 5 shows that an order of magnitude increase in the exchange current density is predicted for a graphene-supported nanocluster compared with an unsupported Ni catalyst. Additional improvement occurs when doping the graphene support with nitrogen, where having nitrogen at the edge of nanocluster (that is, directly interacting with surface Ni atoms) appears to be more effective than having nitrogen at the centre of the nanoparticle. Relating this discovery to the current experimental procedure, it is expected that edge nitrogen is predominant in the experimental material. This is because nitrogen doping is performed after the Ni nanoparticle was grown from the CNT surface. Thus, it is expected that only sites at the edge of the nanoparticle are available for nitrogen doping, and that lattice diffusion is the only mechanism by which nitrogen could be transported to the centre of the nanoparticle.

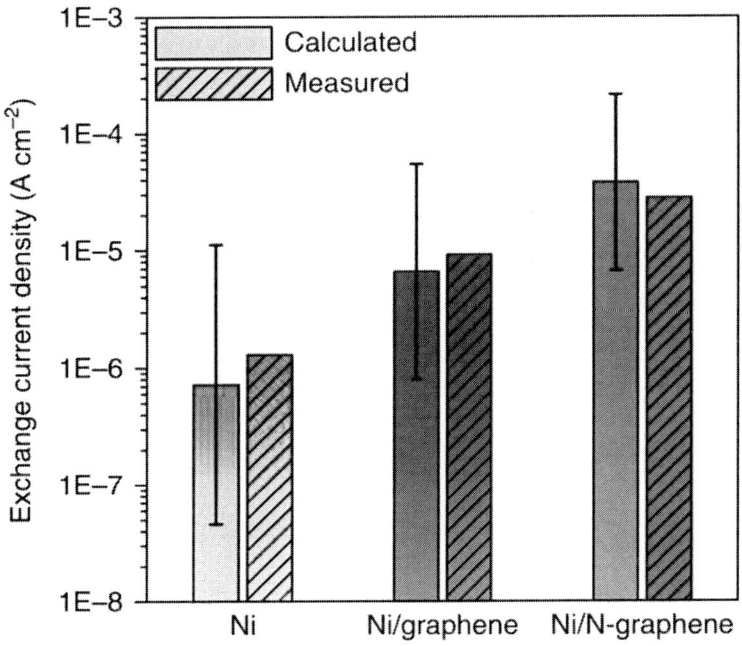

Figure 5. Comparison of calculated exchange current densities to measured values. Unpatterned bars are the calculated exchange current densities and patterned bars are the measured values. The calculated exchange current density of Ni/N_e-graphene is shown for Ni/N-graphene. Error bars are 75% confidence intervals resulting from the regression of the volcano relationship in Supplementary Equation 2.

To confirm that the Ni_{13} is sufficiently large to capture the metal-support interaction of the experimental system, we performed calculations on a larger Ni_{37} nanoparticle (Supplementary Note 2). The *d*-orbitals of the Ni_{13} cluster correspond well to the larger Ni_{37} model (Supplementary Fig. 12). The difference in the calculated *d*-band centre is only 0.01 eV for the two nanoparticles, which indicates that the Ni_{13} is a good approximation for larger nanoparticles both in terms of its adsorption and electronic properties. The electronic effect of the N-dopant was found localized to the adsorption sites of the nearby Ni atoms (Supplementary Fig. 13 andn Supplementary Table 4). Our findings suggest that even higher HOR activity could be obtained by making smaller Ni nanoparticles because of a larger fraction of adsorption sites that could be affected electronically by the N-dopant.

In summary, Ni/N-CNT as a highly active PGM-free HOR catalyst in alkaline condition was reported in this paper. The promoted HOR activity is attributed to the synergetic effect of the edge N atom in the CNT and Ni. The theoretical investigation indicated that the Ni nanoparticle has more sites with an optimal hydrogen-binding energy because of both a geometric and electronic interaction with the edge N atoms. Relative to Ni nanoparticles, the mass activity and exchange current density of Ni/N-CNT increases by a factor of 33 and 21, respectively. These results show that Ni/N-CNT is promising to be applied as the anode catalyst for HEMFCs.

2. METHODS

2.1 Synthesis of nickel-based composite catalyst

In a typical synthesis of Ni/N-CNT, mildly oxidized MWCNTs (25 mg, 10–20 nm in diameter, Shenzhen Nanotech Port Co. Ltd., mildly oxidized through the modified Hummers method[50]) were dispersed in deionized water (10 ml) assisted by sonication for 30 min. After that, $NiSO_4 \cdot 6H_2O$ (0.262 g, 1 mmol) was added to the MWCNT suspension and sonicated for another 30 min. The suspension was transferred to a flask placed in an ice-water bath. Then ice-cold $NaBH_4$ aqueous solution (1 wt %, 10 ml, ~2.5 mmol) was added dropwise. After reaction at 0 °C for 2 h, the suspension was transferred to a 45-ml Teflon-lined stainless steel autoclave. Ammonium hydroxide solution (28–30% NH_3 basis, 5 ml) and hydrazine hydrate solution (78–82%, 2 ml) were added, and then the autoclave was sealed and subjected to a hydrothermal treatment at

150 °C for 4 h. The final product was collected by centrifugation at 8,000 r.p.m. for 10 min, washed with water and then redispersed in water (10 ml).

Ni and Ni/CNT were synthesized by similar procedures. Ni was synthesized without adding oxidized MWCNT and the pH was adjusted to 12 by ammonium hydroxide before the hydrothermal treatment. Ni/CNT was synthesized without adding ammonium hydroxide and hydrazine.

2.2 Physical characterization

The SEM was performed on a JEOL JSM-7400F SEM working at an acceleration voltage of 3 kV. The TEM, HRTEM and EDS were performed on a JEOL JEM-3010 TEM operating at 300 kV equipped with EDS system. The X-ray diffraction patterns were collected on a Bruker D8 Discovery diffractometer with Cu Kα radiation (λ=0.15418, nm) operating at 40 kV and 40 mA. The TGA was performed on a Mettler Toledo TGA/DSA 1 STAR^{e} System under air flow (60 ml min^{-1}) with a heating rate of 5 °C min^{-1}. The XPS spectra were recorded on Thermo Fisher ESCALAB 250Xi XPS system with a monochromatic Al Kα X-ray source. The survey scans were conducted at a pass energy of 100 eV using a step size of 1 eV, and the high-revolution scans were conducted at a pass energy of 50 eV using a step size of 0.05 eV.

2.3 Electrochemical tests

The electrodes were prepared by casting catalyst inks on glassy carbon electrodes (5 mm in diameter, Pine Instruments, polished to a mirror-finishing with 0.05 µm alumina). The catalyst inks were prepared by dispersing catalyst in water with 0.05 wt % Nafion at a concentration of 2.5 mg_{Ni} ml^{-1}. The catalyst ink (20 µl) was deposited on a glassy carbon electrode and dried in vacuum at room temperature, resulted in a metal loading of 0.25 mg_{Ni} cm^{-2}. Electrochemical studies were carried out in a standard three-electrode system controlled with a multichannel potentiostat (Princeton Applied Research). As-prepared thin film electrode, which served as the working electrode, was mounted on a rotator and was immersed into 0.1 M KOH solution. An Ag/AgCl electrode with double junctions was used as the reference electrode. A

Pt wire was used as the counterelectrode, which has an outer glass tube with a ceramic fritz to prevent the contamination of Pt. All potentials reported in this paper are referenced to the reversible hydrogen electrode potential. All the polarization curves were corrected for solution resistance, which was measured using AC-impedance spectroscopy from 200 kHz to 100 mHz and a voltage perturbation of 10 mV.

2.4 Computational methodology

Periodic DFT calculations were performed using the Vienna *ab initio* software package (VASP, version 5.3.2)[51]. The optPBE-vdW exchange-correlation functional[52] was selected because of its qualitatively accurate representation of the potential energy surface of a Ni monolayer and graphene system[53]. The core electrons were represented with the projector-augmented wavefunction (PAW) method[54, 55], and a plane-wave cutoff of 400 eV was used for the valence electrons. The Methfessel–Paxton method of electron smearing was used with a smearing width of 0.1 eV (ref. 56). All geometry optimizations were performed using the conjugate gradient algorithm[57] as implemented in VASP. The forces and energies were converged to 0.05 eV $Å^{-1}$ and 10^{-4} eV, respectively. All calculations were spin-polarized because of the magnetic moment of Ni. For the Ni nanocluster systems, all calculations were performed at the Γ-point. A 5 × 5 graphene unit cell was used as the support. In the *z* direction, a vacuum layer of 15 Å was included. This unit cell was chosen to minimize interaction of the Ni_{13} nanocluster with its periodic image while still being computationally tractable. For the single metal surfaces used to establish the hybrid activity model, a 4 × 4 unit cell was used with a 3 × 3 × 1 Monkhorst–Pack *k*-point sampling of the Brillouin zone[58].

REFERENCES

1. Bockris, J. O. M. Hydrogen economy in the future. *Int. J. Hydrogen Energy* **24**, 1–15(1999).

2. Bockris, J. O. M. The hydrogen economy: its history. *Int. J. Hydrogen Energy* **38**,2579–2588 (2013).

3. Gasteiger, H. A. & Markovic, N. M. Just a dream-or future reality? *Science* **324**, 48–49(2009).

4. Gasteiger, H. A., Kocha, S. S., Sompalli, B. & Wagner, F. T. Activity benchmarks and requirements for Pt, Pt-alloy, and non-Pt oxygen reduction catalysts for PEMFCs. *Appl. Catal. B* **56**, 9–35 (2005).

5. Steinbach, A. *DOE Hydrogen and Fuel Cells Program 2015 Annual Progress Report.* Report no. FC104 (US Department of Energy, Washington, DC, USA, 2015).

6. Varcoe, J. R. & Slade, R. C. T. Prospects for alkaline anion-exchange membranes in low temperature fuel cells. *Fuel Cells* **5**, 187–200 (2005).

7. Chung, H. T., Won, J. H. & Zelenay, P. Active and stable carbon nanotube/nanoparticle composite electrocatalyst for oxygen reduction. *Nat. Commun.* **4**, 1922 (2013).

8. Wu, G. & Zelenay, P. Nanostructured Nonprecious Metal Catalysts for Oxygen Reduction Reaction. *Acc. Chem. Res.* **46**, 1878–1889 (2013).

9. Piana, M. *et al.* H-2/air alkaline membrane fuel cell performance and durability, using novel ionomer and non-platinum group metal cathode catalyst. *J. Power Sources* **195**, 5875–5881(2010).

10. Sa, Y. J. *et al.* Carbon nanotubes/heteroatom-doped carbon core-sheath nanostructures as highly active, metal-free oxygen reduction electrocatalysts for alkaline fuel cells. *Angew. Chem. Int. Ed. Engl.* **53**, 4102–4106 (2014).

11. Durst, J. *et al.* New insights into the electrochemical hydrogen oxidation and evolution reaction mechanism. *Energy Environ. Sci.* **7**, 2255–2260 (2014).

12. Strmcnik, D. *et al.* Improving the hydrogen oxidation reaction rate by promotion of hydroxyl adsorption. *Nat. Chem.* **5**, 300–306 (2013).

13. Sheng, W., Gasteiger, H. A. & Shao-Horn, Y. Hydrogen oxidation and evolution reaction kinetics on platinum: acid vs alkaline electrolytes. *J. Electrochem. Soc.* **157**,B1529–B1536 (2010).

14. Wang, Y. *et al.* Pt-Ru catalyzed hydrogen oxidation in alkaline media: oxophilic effect or electronic effect? *Energy Environ. Sci.* **8**, 177–181 (2015).

15. Nocera, D. G. The artificial leaf. *Acc. Chem. Res.* **45**, 767–776 (2012).

16. Gong, M. *et al.* Nanoscale nickel oxide/nickel heterostructures for active hydrogen evolution electrocatalysis. *Nat. Commun.* **5**, 4695 (2014).

17. McCrory, C. C. L. *et al.* Benchmarking hydrogen evolving reaction and oxygen evolving reaction electrocatalysts for solar water splitting devices. *J. Am. Chem. Soc.* **137**,4347–4357 (2015).

18. Kenjo, T. Chromium-doped Raney-nickel catalyst for hydrogen electrodes in alkaline fuel-cells. *J. Electrochem. Soc.* **132**, 383–386 (1985).

19. Kiros, Y. & Schwartz, S. Long-term hydrogen oxidation catalysts in alkaline fuel cells. *J. Power Sources* **87**, 101–105 (2000).

20. Mund, K., Richter, G. & Sturm, F. V. Titanium-containing Raney-nickel catalyst for hydrogen electrodes in alkaline fuel-cell systems. *J. Electrochem. Soc.* **124**, 1–6 (1977).

21. Lu, S., Pan, J., Huang, A., Zhuang, L. & Lu, J. Alkaline polymer electrolyte fuel cells completely free from noble metal catalysts. *Proc. Natl Acad. Sci. USA* **105**, 20611–20614(2008).

22. Sheng, W. *et al.* Non-precious metal electrocatalysts with high activity for hydrogen oxidation reaction in alkaline electrolytes. *Energy Environ. Sci.* **7**, 1719–1724 (2014).

23. Wood, K. N., O'Hayre, R. & Pylypenko, S. Recent progress on nitrogen/carbon structures designed for use in energy and sustainability applications. *Energy Environ. Sci.* **7**,1212–1249 (2014).

24. Yang, Z. X., Xia, Y. D. & Mokaya, R. Aligned N-doped carbon nanotube bundles prepared via CVD using zeolite substrates. *Chem. Mater.* **17**, 4502–4508 (2005).

25. Wang, H., Maiyalagan, T. & Wang, X. Review on recent progress in nitrogen-doped graphene: synthesis, characterization, and its potential applications. *ACS Catal.* **2**,781–794 (2012).

26. Markovic, N. M., Sarraf, S. T., Gasteiger, H. A. & Ross, P. N. Hydrogen electrochemistry on platinum low-index single-crystal surfaces in alkaline solution. *J. Chem. Soc. Faraday Trans.* **92**, 3719–3725 (1996).

27. Ewe, H., Justi, E. & Schmitt, A. Structure and properties of Raney-nickel catalysts with alloying for alkaline fuel-cells. *Electrochim. Acta* **19**, 799–808 (1974).

28. Ewe, H., Justi, E. & Schmitt, A. Studies of preservation of Raney-nickel-mixed catalysts.*Energy Convers.* **14**, 35–41 (1975).

29. Jenseit, W., Khalil, A. & Wendt, H. Meterial properties and processing in the production of fuel-cell components. 1. Hydrogen anodes from Raney-nickel for lighweight alkaline fuel-cells. *J. Appl. Electrochem.* **20**, 893–900 (1990).

30. AlSaleh, M. A., Gultekin, S., AlZakri, A. S. & Khan, A. A. A. Steady state performance of copper impregnated Ni/PTFE gas diffusion electrode in alkaline fuel cell. *Int. J. Hydrogen Energy* **21**, 657–661 (1996).

31. SleemUrRahman, M. A., AlSaleh, M. A., AlZakri, A. S. & Gultekin, S. Preparation of Raney-Ni gas diffusion electrode by filtration method for alkaline fuel cells. *J. Appl. Electrochem.* **27**, 215–220 (1997).

32. Al-Saleh, M. A., Sleem, Ur, R., Kareemuddin, S. & Al-Zakri, A. S. Novel methods of stabilization of Raney-Nickel catalyst for fuel-cell electrodes. *J. Power Sources* **72**,159–164 (1998).

33. Machado, S. A. S. & Avaca, L. A. The hydrogen evolution reaction on nickel surfaces stabilized by H-absorption. *Electrochim. Acta* **39**, 1385–1391 (1994).

34. Grden, M., Alsabet, M. & Jerkiewicz, G. Surface science and electrochemical analysis of nickel foams. *ACS Appl. Mater. Interfaces* **4**, 3012–3021 (2012).

35. Henning, S., Herranz, J. & Gasteiger, H. A. Bulk-palladium and palladium-on-gold electrocatalysts for the oxidation of hydrogen in alkaline electrolyte. *J. Electrochem. Soc.* **162**, F178–F189 (2015).

36. Wulff, G. On the question of speed of growth and dissolution of crystal surfaces. *Z. Krystallogr. Mineral.* **34**, 449–530 (1901).

37. Miehle, W., Kandler, O., Leisner, T. & Echt, O. Mass-spectrometric evidence for icosahedral structrue in large rare-gas clusters -Ar, Kr, Xe. *J. Chem. Phys.* **91**, 5940–5952(1989).

38. Popov, V. N. Curvature effects on the structural, electronic and optical properties of isolated single-walled carbon nanotubes within a symmetry-adapted non-orthogonal tight-binding model. *New J. Phys.* **6**, 17 (2004).

39. Trasatti, S. Work Function, electronegativity, and electrochemical behavior of metals. 3. Electrolytic hydrogen evolution in acid solutions. *J. Electroanal. Chem.* **39**, 163–184 (1972).

40. Jerkiewicz, G. & Zolfaghari, A. Determination of the energy of the metal–underpotential-deposited hydrogen bond for rhodium electrodes. *J. Phys. Chem.* **100**, 8454–8461 (1996).

41. Jerkiewicz, G. Hydrogen sorption AT/IN electrodes. *Prog. Surf. Sci.* **57**, 95–186 (1998).

42. Conway, B. E. & Tilak, B. V. Interfacial processes involving electrocatalytic evolution and oxidation of H-2, and the role of chemisorbed H. *Electrochim. Acta* **47**, 3571–3594 (2002).

43. Norskov, J. K. *et al.* Trends in the exchange current for hydrogen evolution. *J. Electrochem. Soc.* **152**, J23–J26 (2005).

44. Sheng, W., Myint, M., Chen, J. G. & Yan, Y. Correlating the hydrogen evolution reaction activity in alkaline electrolytes with the hydrogen binding energy on monometallic surfaces. *Energy Environ. Sci.* **6**, 1509–1512 (2013).

45. Sheng, W. *et al.* Correlating hydrogen oxidation and evolution activity on platinum at different pH with measured hydrogen binding energy. *Nat. Commun.* **6**, 5848 (2015).

46. Peela, N. R., Zheng, W., Lee, I. C., Karim, A. M. & Vlachos, D. G. Core-shell nanocatalyst design by combining high-throughput experiments and first-principles simulations.*Chemcatchem* **5**, 3712–3718 (2013).

47. Jenness, G. R. & Schmidt, J. R. Unraveling the role of metal-support interactions in heterogeneous catalysis: oxygenate selectivity in fischer-tropsch synthesis. *ACS Catal.* **3**,2881–2890 (2013).

48. Hammer, B. & Norskov, J.K. in *Advances in Catalysis, Vol. 45* eds Gates B. C., Knozinger H. 71–129Elsevier (2000).

49. Gorzkowski, M. T. & Lewera, A. Probing the limits of d-band center theory: electronic and electrocatalytic properties of Pd-shell-Pt-core nanoparticles. *J. Phys. Chem. C* **119**, 18389–18395 (2015).

50. Liang, Y. *et al.* Oxygen reduction electrocatalyst based on strongly coupled cobalt oxide nanocrystals and carbon nanotubes. *J. Am. Chem. Soc.* **134**, 15849–15857 (2012).

51. Kresse, G. & Furthmuller, J. Efficiency of *ab-initio* total energy calculations for metals and semiconductors using a plane-wave basis set. *Comput. Mater. Sci.* **6**, 15–50 (1996).

52. Klimes, J., Bowler, D. R. & Michaelides, A. Chemical accuracy for the van der Waals density functional. *J. Phys. Condens. Matter* **22**, 022201 (2010).

53. Wellendorff, J. *et al.* Density functionals for surface science: exchange-correlation model development with Bayesian error estimation. *Phys. Rev. B* **85**, 235149 (2012).

54. Blochl, P. E. Projector augmented-wave method. *Phys. Rev. B* **50**, 17953–17979 (1994).

55. Kresse, G. & Joubert, D. From ultrasoft pseudopotentials to the projector augmented-wave method. *Phys. Rev. B* **59**, 1758–1775 (1999).

56. Methfessel, M. & Paxton, A. T. High-precision sampling for Brillouin-zone intergration in metals. *Phys. Rev. B* **40**, 3616–3621 (1989).

57. Teter, M. P., Payne, M. C. & Allan, D. C. Solution of Schrodinger-equation for large systems. *Phys. Rev. B* **40**, 12255–12263 (1989).

3

CHARACTERIZATION OF LASER-INDUCED DEFECTS AND MODIFICATION IN CARBON NANOTUBES BY RAMAN SPECTROSCOPY

Masaru Tachibana

Department of Nanosystem Science, Yokohama City University, Japan

1. INTRODUCTION

Defects in single-wall carbon nanotubes (SWCNTs) have a great influence on their physical properties. In real SWCNTs, various types of defects such as vacancies, Stone–Wales defects, adatoms, or H–C complex are contained as shown in Figure 1. Such defects can be introduced at the stage of SWCNT growth or later on during device or composite production. They can be also created deliberately by chemical treatment or by irradiation with electron, ion, or laser light. Understanding the properties of such defects in SWCNTs is important for improving SWCNT growth methods, tailoring their physical properties, and controlling the irradiation-induced damages.

The irradiation with electron, ion, or laser light has been widely used for the studies not only on the properties of defects but also on the modification of CNTs. The nice reviews on electron and ion irradiation-induced effects in CNTs have been already published by Krasheninnikov *et al.* [1,2]. However, there are very few reviews on laser-induced defects. The laser irradiation can lead to the heating followed by burning. Such laser irradiation also gives rise to interesting effects such as the production of defects and the modification of CNTs [3-7]. This chapter presents our recent studies [8-11] on laser-induced effects in SWCNTs.

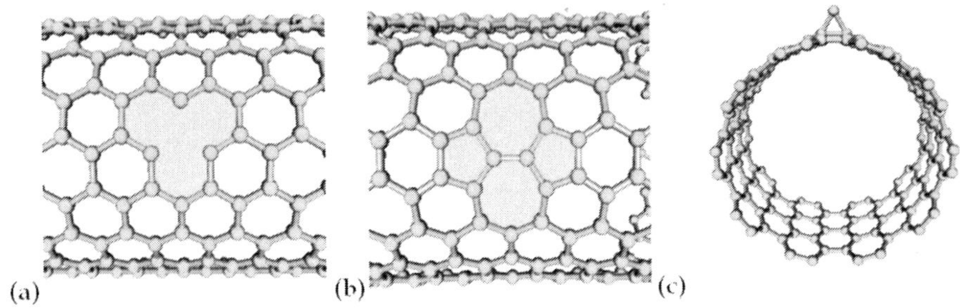

Figure 1. Schematic figures of typical defects such as (a) vacancy, (b) Stone-Wales defect, and (c) adatom in SWCNTs.

Resonant Raman spectroscopy is one of the most powerful tools for characterizing structural and electronic properties of SWCNTs [12]. In the Raman spectra, defect-induced phonon mode so-called D band is observed at around 1350 cm^{-1}. The intensity of the D band can be enhanced as the number of defects is increased in the SWCNT. Therefore, the D band has been used for the assessment of imperfection of SWCNTs and the understanding of the properties of their defects. Further finding Raman bands associated with defects can lead the Raman spectroscopy to a more effective tool for the characterization of defects.

This chapter presents our recent studies [8-11]on the characterization of laser-induced defects and modification in SWCNTs by Raman spectroscopy. This chapter consists of four parts as mentioned below: (1) Thermal relaxation of laser-induced defects in SWCNTs, (2) Phonon control in metallic SWCNTs by laser–induced defects, (3) Fine structure of D band related to laser-induced defects in SWCNTs, and (4) Formation of *trans*-polyacetylene from SWCNTs by laser irradiation.

2. RAMAN SPECTRA OF SWCNTS

The resonant Raman spectra of SWCNTs include two main features: a radial breathing mode (RBM) observed in the range of 50–350 cm^{-1} and a tangential mode (the so-called G band) observed in the range of 1450–1650 cm^{-1} [12].

The RBM is a signature for the presence of SWCNTs, and is observed as a peak or a multi-peak feature. In the RBM, as suggested by its name, all the C atoms are vibrating in the radial direction with the same phase, as if the tube are breathing. The atomic motion does not break the tube symmetry, that is, the RBM is a totally symmetric (A_1) mode. Since this particular vibrational mode only occurs in SWCNTs, it is used to distinguish SWCNTs from other sp^2 carbons such as graphite and graphene.

A very important characteristic is the RBM frequency (ω_{RBM}) dependence on the tube diameter (d_t):$\omega_{RBM} \infty 1/d_t$ [13]. Most of the RBM experiment results in the literature have been fitted with the relation:

$$\omega_{\mathrm{RBM}} = A / d_{\mathrm{t}} + B \tag{1}$$

with values for the parameters A and B varying widely from paper to paper. Thus, the RBM can give an easy and quick determination of the tube diameter. In addition, it is of important that the RBM peak intensity is a function of excitation energy. The RBM is intense when the incident light or the scattered light is in resonance with the SWCNT optical transition energies.

The G band (where the notation G comes from graphite) is related to the in plane C–C bond stretching mode in graphite and graphene. The G band is the Raman signature for all the sp^2 carbon materials, and is observed as a peak or multi-peak feature. The G band in SWCNTs is a more complex spectral feature. Due to the folding of graphene sheet into the SWCNT and the symmetry breaking effects associated with the nanotube curvature, G band splits into G^+ and G^-, which are related to atomic vibrations preferencially along (LO) and perpendicular (TO) to the tube (folding) axis, respectively, for semiconducting SWCNT. For metallic tubes, electron-phonon coupling softens the LO modes, so that G^+ and G^- are actually with TO and LO modes, respectively.

The G- peak for metallic tubes is fitted by asymmetric and broad Breit-Wigner-Fano (BWF) line:

$$I(\omega)=I_0\frac{[1+(\omega-\omega_{BWF})/q\Gamma]^2}{1+[(\omega-\omega_{BWF})/\Gamma]^2} \quad (2)$$

where I_0, ω_{BWF}, Γ, and $1/q$ are intensity, renormalized frequency, broadening parameter, and the asymmetric parameter, respectively [14]. Note that $\omega_{BWF}$and 2Γ are called the peak frequency and full width at half maximum (FWHM) of the BWF line, respectively. As clarified from Equation (2), $1/q$determines the departure of the line shape from a symmetric Lorentzian line and, therefore, $|1/q|$ is a measure of the degree of the BWF coupling. Thus the asymmetric and broad BWF line is commonly used to distinguish metallic SWCNTs from semiconducting SWCNTs as fitted with symmetric Lorentzian lines.

Actually, due to the symmetry breaking effects associated with the nanotube curvature, G band in SWCNT generates up to six Raman-allowed G-band peaks corresponding to two totally symmetric A_1modes, two E_1 modes and two E_2 symmetry modes. Three of each exhibit LO or TO-like vibration. Due to the delocalization effect and special resonance conditions, the A_1 modes usually dominante the G-band spectra.

In addtion, in the Raman spectra in SWCNTs, defect-induced phonon mode so-called D band is often observed at around 1350 cm^{-1} [12].The D band is a Raman signature of disorder in sp^2 carbons materials. The intensity of the D band can be enhanced as the number of defects is increased in the SWCNT. The D band has been used for the assessment of imperfection of SWCNTs and the understanding of the properties of their defects.

3. THERMAL RELAXATION OF LASER-INDUCED DEFECTS IN SWCNTS

3.1. Laser Irradiation For Swcnts Synthesized By Electric Arc-Discharge Method

SWCNTs synthesized by an electric arc-discharge method were used for laser irradition experiments. As-grown SWCNTs were purified by heating at 350°C for 90 min in air. A suspension of purified SWNTs in ethanol was prepared by ultrasonication. By drop-coating and air-drying the suspension, a SWCNT thin film was formed on a quartz substrate. The SWCNT film was irradiated with a 248 nm (~5.0 eV) pulsed KrF excimer laser in air. The irradiation fluence was approximately 3 J/(cm^2·pulse). The irradiation pulse number was selected to be only one because more than two pulses led to the breakdown of the SWCNTs.

3.2. Change In D Band By Laser Irradiaton And Thermal Annealing

Figure 2 shows D and G bands in the Raman spectra for non-irradiated SWCNTs, irradiated SWNTs with a 248 nm pulsed excimer laser of 3 J/(cm^2·pulse) in air, and annealed SWCNTs at 673 K in a vacuum of 1 Pa for 240 min after the laser irradiation. The inset shows the close up of the D band. All of the spectra were normalized to the maximum intensity of the G band. Note that the Raman excitation was provided with a 532 nm (~2.33 eV) of a Nd:YVO4 laser where the laser power level in a focal spot of 1 μm in diameter on the sample was kept below 0.1 mW to prevent overheating the sample.

It is found that the D band intensity significantly changes by laser irradiation and thermal annealing, while the spectral feature of the G band remains almost unchanged. For more clarifying the change in D band intensity, the relative intensity of the D band main peak at 1346 cm^{-1} to the G band main peak at 1593 cm^{-1} was defined as I_D/I_G. For non-irradiated SWCNTs as shown in Fig. 2(a), the $I_D/I_G$is 0.009. This value is very small and exhibits high quality SWCNTs. The I_D/I_Gsignificantly increases due to laser irradiation as seen in Fig. 2(b). The $I_D/I_G$for the irradiated SWCNTs is 0.08, which is about ten times as much as that for the non-irradiated ones. Furthermore, when the irradiated SWNTs are thermally annealed at 673 K for 240 min, the I_D/I_Gdecreases and approaches that for non-irradiated ones as shown in Fig. 2(c). On the other hand, all peak frequencies, relative intensities, and FWHMs of RBM, D and G bands except for the intensity of the D band exhibit no significant change due to laser irradiation and thermal annealing. These spectral features mean that defects were successfully produced by the laser irradiation and relaxed by annealing with keeping the tubular structure of SWCNTs.

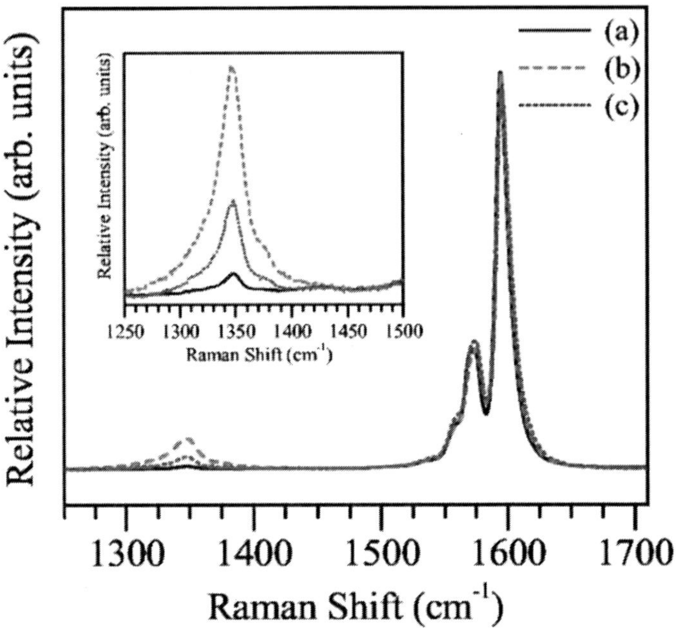

Figure 2. D and G bands in the Raman spectra of (a) pristine non-irradiated SWCNTs, (b) irradiated SWCNTs with a 248 nm pulsed excimer laser, and (c) annealed SWCNTs at 673 K for 240 min after the laser irradiation. The inset shows the close up of the D band. All of the spectra are normalized to the maximum intensity of the G band. [**8**]

Let us consider the formation process of the laser-induced defects in SWCNTs. The knock-on energy of carbon atom into the direction perpendicular to the tube surface for an isolated SWCNT with a diameter over 1 nm is estimated to be 15–17 eV [15]. This energy is much higher than the irradiation energy of 248 nm (~5.0 eV) used in this experiment. This means that the formation of the laser-induced defects in SWCNTs would not be due to the physical knock-on phenomena. However, the increase of D band intensity related to the formation of defects clearly occurs for SWCNTs irradiated in air. The degree of the increase of D band intensity is much higher than those for SWCNTs irradiated in vacuum and Ar atmosphere. Therefore, the formation of the laser-induced defects in SWCNTs can be attributed to the chemical reaction with O_2 and H_2O in air by laser heating.

3.3. Analysis Of Recovery Of D Band By Themal Annealing

To examine the thermal relaxation of the laser-induced defects in SWCNTs, the time evolution of the relative intensity of D band in the irradiated samples at various annealing temperatures from 296 to 698 K was measured in the range of annealing times of 0 to 240 min. The typical annealing time evolution of the relative intensity of the D band at 573, 673, and 698 K are shown in Figure 3. Note that the *R* of the ordinate in the figure indicates the I_D/I_G normalized by that at t=0, i.e., before annealing for irradiated SWCNTs.

As shown in Fig. 3, the *R* decreases with increasing annealing time. For 573 K, the *R* gradually decreases and reaches 0.8 at 240 min. For 673 K, the *R* quickly decreases and reaches 0.6 at 20 min, and then slowly decreases and reaches 0.45 at 240 min. For 698 K, the *R* more quickly decreases and reaches 0.4 at 20 min, and then slowly decreases and reaches 0.3 at 240 min. On the other hand, no significant change in the *R* is observed for 296 K. These results indicate that the behavior of the *R* with annealing time strongly depends on annealing temperature. The feature of the dependence of the *R* on annealing time is clearly observed, especially at annealing temperatures of 673 and 698 K as shown inFigs. 3(b) and 3(c), respectively. From these behaviors, it is predicted that the thermal relaxation of laser-induced defects in SWCNTs includes two processes that have fast and slow rates.

As the previous analysis of the thermal relaxation kinetics of defects for graphite [16], the dependence of the *R* on annealing time for SWCNTs as shown in Fig. 3 was analyzed on the following three assumptions: (i) the I_D/I_G, or *R* is proportional to the density of defects in SWCNTs, (ii) some of the defects and others are annihilated by the fast and slow relaxation processes, respectively, and (iii) both processes occur independently. According to these assumptions, the annealing time evolution of the *R*can be expressed as

$$R(t) = A\exp(-k_{I}t) + (1 - A)\exp(-k_{II}t) \tag{3}$$

where *A* is the ratio of the defects relaxed by the fast relaxation process to the total defects in SWCNTs, k_I and k_{II} are the rate constants for the fast and slow relaxation processes, respectively, and*t* is annealing time. So the first and second terms on the right-hand side of Equation (3) correspond to the fast and slow relaxation processes, respectively. The annealing time evolution of the *R* as shown inFig. 3 was fitted with Eq.

(3). The fitted curves are indicated as solid lines in Fig. 3. As a result, the rate constants of k_{I} and k_{II} for the fast and slow relaxation processes, respectively, are determined as shown in Table I. For 296 K, k_{I} and k_{II} are estimated to be quite small ($<10^{-6}$) since no clear change in R was observed. The value of A runs from 0.23 through 0.59. As presented in Table I, both k_{I} and k_{II} show the strong temperature dependence.

3.4. Activation Energies Of Thermal Relaxation Of Laser-Induced Defects

Arrhenius plots of thermal relaxation rate constants of $k_{I}\,t$ and $k_{II}\,t$in Table I are shown in Figure 4. From the slopes in Fig. 4, the activation energies for the fast and slow relaxation processes are determined to be 0.4 and 0.7 eV, respectively. According to the simulation on defects in SWCNTs, the activation energy of the vacancy migration along the tube axis is calculated to be about 1 eV [17]. This value is close to that for the slow relaxation process as determined experimentally above. Such vacancy migration would also lead to the annihilation of the vacancies at the nanotube ends. Therefore, it is concluded that the slow relaxation process observed in this experiment corresponds to the vacancy migration along the tube axis.

On the other hand, two such relaxation processes have also been reported for graphite irradiated with He+ ions [16]. According to the report, the fast and slow relaxation processes correspond to the vacancy-interstitial recombination and vacancy migration in the graphene plane, respectively, which have the activation energies of 0.89 eV and 1.8 eV. This suggests that the slow relaxation process corresponds to the vacancy migration in both SWCNT and graphite. The activation energy of 0.7 eV of the vacancy migration for SWCNTs as deterimined experimentally above is much smaller than 1.8 eV for graphite. This smaller value for SWCNTs would be due to the curvature effect of nanotube that breaks the trigonal symmetry of a perfect graphene sheet. It is expected that, due to the curvature effect of nanotube, the activation energy of the vacancy-interstitial recombination for SWCNTs is also smaller than 0.89 eV for graphite. Thus, it is suggested that the fast relaxation process with the activation energy of 0.4 eV as determined in this experiment corresponds to the vacancy-interstitial recombination in SWCNTs.

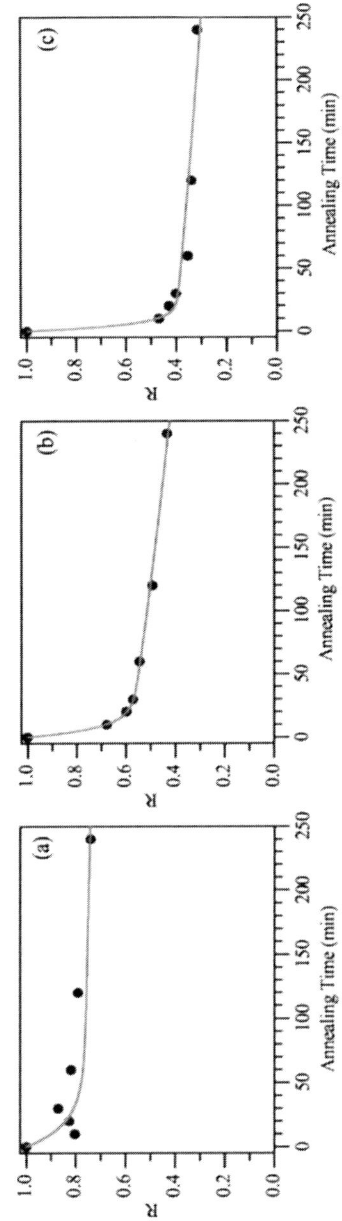

Figure 3. Time evolution of the R at annealing temperatures of (a) 573, (b) 673, and (c) 698 K for SWCNTs irradiated with a 248 nm pulsed excimer laser. Note that the R indicates the ID/ IG normalized by that at t=0, i.e., before annealing for the irradiated SWCNTs. Solid curves were obtained by the fitting analysis with **Eq. (3)**. [8]

Table 1. Thermal relaxation rate constants of k_{I} and k_{II} for the fast and slow processes, respectively, at annealing temperatures from 296 to 698 K for SWCNTs irradiated with a 248 nm pulsed excimer laser.[**8**]

Temperature (K)	k_I (s^{-1})	k_{II} (s^{-1})
296	$< 10^{-6}$	$< 10^{-6}$
573	9.5×10^{-4}	2.5×10^{-6}
623	1.5×10^{-3}	5.6×10^{-6}
648	2.9×10^{-3}	1.6×10^{-5}
673	2.4×10^{-3}	2.2×10^{-5}
698	3.5×10^{-3}	2.0×10^{-5}

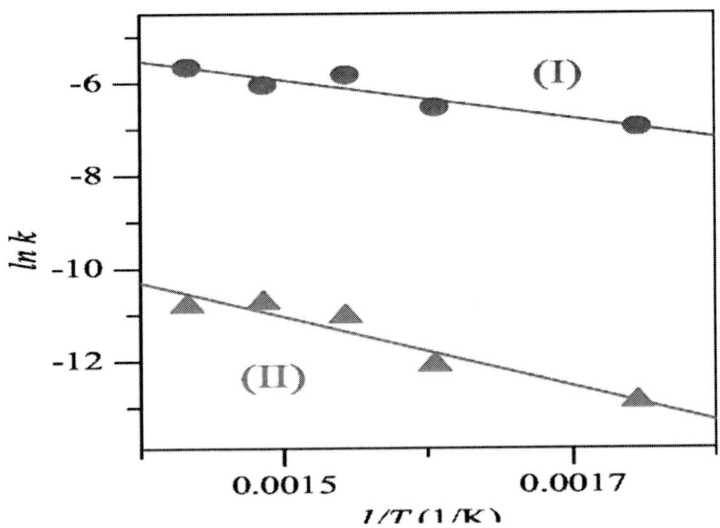

Figure 4. Arrhenius plots of the thermal relaxation rate constants of k_{I}(circles) and k_{II} (triangles) for the fast and slow processes, respectively, for SWCNTs irradiated with a 248 nm pulsed excimer laser, where$T$ is the annealing temperature. [**8**]

In summary, laser-induced defects in SWCNTs can be introduced by the irradiation with a 248 nm pulsed excimer laser. The formation of defects might be related to thermal oxidation and burning by laser heating. Such laser-induced defects are thermally relaxed with two processes with fast and slow rates. The two relaxation processes show the strong temperature dependence. The activation energies of the fast and slow relaxation processes are determined to be 0.4 and 0.7 eV, respectively. These processes can correspond to vacancy-interstitial recombination and vacancy migration along the tube axis. Such relaxation processes with fast and slow rates for SWCNTs are similar to those for graphite irradiated with He+ ions. However, their activation energies for SWCNTs are smaller than those for graphite. The smaller activation energies for SWCNTs would be due to the effect of curvature of nanotube.

4. PHONON CONTROL IN METALLIC SWCNTS BY LASER-INDUCED DEFECTS

4.1. Change In G Band For Metallic Nanotubes By Laser Irradiation And Thermal Annealing

SWCNT samples after laser irradiation and themal annealing were prepared by similar procedure as described in 3.1. Figure 5 show D and G bands in Raman spectra for (a) pristine SWCNTs, (b) laser-irradiated SWCNT swith a 248 nm (~5.0 eV) pulsed KrF excimer laser of approximately 5 J/(cm^2·pulse), and (c) annealed SWNTs at 400 °C in a vacuum of ~9 Pa for 60 min after the irradiation. Note that the spectra excitation was provided with a 632.8 nm (~1.96 eV) of a He–Ne laser where the laser power level in a focal spot of 1 μm in diameter on the sample was kept below 0.1 mW to prevent overheating the samples.

D band was fitted with one Lorentzian line. On the other hand, G band was fitted with two Lorentzian lines and one asymmetric line that is Breit–Wigner–Fano (BWF) line [13] as given by Eq. (2). The values of the fitting parameters are listed in Table 2. In the G band for pristine SWCNTs in Fig. 5(a),one Lorentzian line at 1593 cm^{-1} corresponds to G$^+$ peaks for semiconducting SWCNTs and metallic SWCNTs. Note that G$^+$ peak associated with semiconducting SWCNTs is assumed to overlap with that with metallic ones. The other Lorentzian line at 1567 cm^{-1} corresponds to G$^-$ peak associated with semiconducting SWCNTs. The asymmetric and broad line at 1546 cm^{-1} corresponds to G$^-$ peak associated with metallic SWCNTs, which is largely downshifted, relative to G^+ one. These spectral components of the G band suggest that both semiconducting and metallic SWCNTs are resonant in the Raman spectrum of pristine SWCNTs taken with 1.96 eV in Fig. 5(a).

The D band for pristine SWCNTs was fitted with one Lorentzian curve at 1323 cm^{-1}. It is known that the intensity of the D band increases as the number of defects in SWCNTs is increased. The intensity ratio of D/G^+ peaks is often used as a measure of the defect density in SWNTs. The relative intensity of the D band for pristine SWCNTs is 0.12. This value is quite small, comparable to those of high quality SWCNTs as reported so far. Thus, pristine SWCNTs used in this experiment have high quality or quite low defect density.

A significant change in the Raman spectrum was observed for laser-irradiated SWCNTs, as shown inFig. 5(b). The intensity of the D band increased with the laser irradiation. Note that the frequency and linewidth remained almost unchanged even after the irradiation. These results mean that some specific defects were introduced in SWCNTs by the laser irradiation. The formation of defects might be related to thermal oxidation and burning by laser heating as discussed in 3.2. Moreover, it should be noticed that not only D band but also G band were affected by the laser irradiation. Especially, a significant change was observed for G^- peak associated with metallic tubes. The frequency of the G^- peak was upshifted by 8 cm^{-1}. Correspondingly, the linewidth of the G^- peak was reduced by 25%. In addition, the intensity ratio of the G^-/G^+ peak was reduced by 25%. On the other hand, the frequency, linewidth, relative intensity of G– peak associated with semiconducting SWCNTs remained almost unchanged even after the laser irradiation. Such behavior in the G^- peak associated with semiconducting SWCNTs is consistent with that in the G^- peak for the same SWCNTs taken with 2.33 eV in which only semiconducting SWNTs are resonant. Thus, the laser-induced defects significantly affect G^- peak associated not with semiconducting SWCNTs but metallic ones.

The G^- peak associated with metallic SWCNTs is due to the electron-phonon coupling as described in 2 [18,19]. The upshift of the frequency, the narrowing of the linewidth, and the reduction in the relative intensity for the G^- peak associated with metallic SWCNTs as seen in Fig. 5(b) imply the breaking of the electron-phonon coupling. Moreover, it should be noticed that D and G bands for the irradiated SWNTs recover the original ones after annealing in a vacuum of ~9 Pa at a sample temperature of 400 °C for 60 min, as shown in Fig. 5(c). As described in 3.4, the laser-induced defects such as vacancies can be annihilated by vacancy-interstitial recombination and vacancy migration to the nanotube end due to the thermal annealing [8]. Such annihilation of vacancies can be responsible for the recovery of the Raman spectral profile. Thus, the electron-phonon coupling can be reversibly controlled by the generation and annihilation of specific defects due to laser irradiation and thermal annealing.

4.2. Change In Rbm For Metallic Nanotubes By Laser Irradiation And Themal Annealing

The change corresponding to that in D and G bands in Fig. 5 was also observed for radial breathing modes (RBMs) at the range of 150–200 cm^{-1} in Raman spectra for pristine SWNTs, laser-irradiated SWCNTs, and annealed SWCNTs after the irradiation, taken with E_{exc} =1.96 eV(λ_{exc}=632.8 nm), as shown in Figure 6. The RBMs are actually composed of a lot of peaks corresponding to various kinds of chiralities or diameters of SWNTs. For simplifying, the RBMs were fitted with five Lorentzian lines. The values of the fitting parameters are listed in Table 3.

The diameters d of SWCNTs resonantly contributing to the Raman spectrum for pristine SWCNTs inFig. 6(a) are estimated to be d=1.44±0.2 nm from the corresponding RBM frequencies ω_{RBM} using the relation $\omega_{RBM}(cm^{-1}) = 234/d(nm) + 10$ [20], which has been found for typical SWCNT bundles. Moreover, according to the (revised) Kataura plot [21,22] with the excitation energy and SWCNT diameters estimated above, it is suggested that both semiconducting and metallic SWCNTs are resonant in the Raman spectrum in Fig. 6(a). Note that one RBM peak at the lowest frequency of 157 cm^{-1} is mainly associated with semiconducting ones whereas other four ones at higher frequencies of 168, 173, 186, and 196 cm^{-1} are associated with metallic ones. This result is consistent with the result of spectral components in G band as discussed in 4.1.

The RBM peaks associated with metallic SWCNTs changed after the irradiation, as shown in Fig. 6(b). Especially, the most intense RBM peak at 173 cm^{-1} drastically decreased with the irradiation. On the other hand, no significant change was observed for only RBM peak at 157 cm^{-1} associated with semiconducting SWCNTs. These results mean that the resonant off for the Raman excitation of 1.96 eV occurs for metallic SWCNTs. This also suggests that the change in the electronic structure for metallic SWCNTs occurs due to the laser-induced defects.

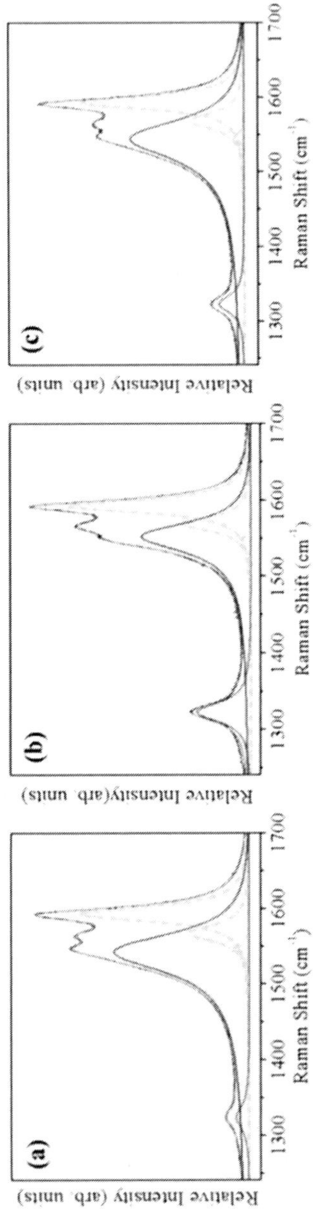

Figure 5. D and *G* bands in Raman spectra for (a) pristine SWNTs, (b) laser-irradiated SWNTs, and (c) annealed SWNTs after the irradiation, taken at E_{exc}=1.96 eV (λ_{exc}=632.8 nm). The *D* band was fittedwith one Lorentzian line. The *G* band was fitted with two Lorentzian (dotted) lines and one BWF (solid) line. **[9]**

Table 2. Peak frequencies (ω), full widths at half maximum (Γ), and relative intensities ($I/I_{G^+(S+M)}$) of Lorentzian lines and Breit-Wigner-Fano lines used to fit *D* and *G* bands for pristine SWCNTs, laser-irradiated SWNTs, and annealed SWCNTs after the irradiation in **Fig. 5**. The relative intensities of the peaks are normalized by G^+ intense peak located at highest frequency. S and M indicate *G* peaks associated with semiconducting and metallic SWCNTs, respectively.

	pristine SWCNTs			laser-irradiated SWCNTs			annealed SWCNTs		
	ω (cm^{-1})	Γ (cm^{-1})	$I/I_{G^+(S+M)}$	ω (cm^{-1})	Γ (cm^{-1})	$I/I_{G^+(S+M)}$	ω (cm^{-1})	Γ (cm^{-1})	$I/I_{G^+(S+M)}$
G^+(S+M)	1593	24	1.0	1592	21	1.0	1593	21	1.0
G^-(S)	1567	26	0.36	1568	23	0.38	1568	27	0.33
G^-(M)	1546	63	0.78	1554	47	0.59	1548	65	0.69
D	1323	38	0.12	1322	31	0.42	1322	28	0.20

Vacancy defects can cause a bandgap opening in metallic SWCNTs due to the breaking of the symmetry [23,24]. Such metal-semiconductor transition has been also experimentally demonstrated by the measurements of electrical properties for metallic SWCNTs with the introduction of defects [25,26]. Therefore, the change in the electronic structure with the bandgap opening can be responsible for the resonant off for metallic SWCNTs. Such change in the electronic structure for metallic SWCNTs due to the laser-induced defects is also consistent with the change in the corresponding G⁻band, i.e., the breaking of the electron-phonon coupling, as discussed in 4.1.

Moreover, as seen in Fig. 6(c), the thermal annealing also leads to the recovery of the RBMs to original ones, as D and G band in Fig. 5. This recovery can be also explained by that in the electronic structure due to the thermal annihilation of laser-induced defects such as vacancies as discussed in 3.4.

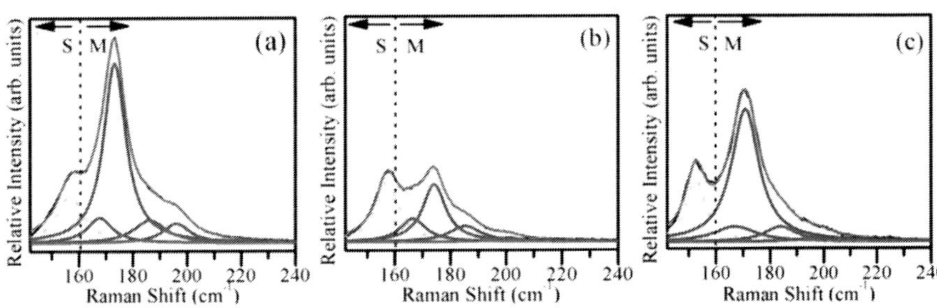

Figure 6. RBMs in Raman spectra for (a) pristine SWCNTs, (b) laser-irradiated SWCNTs, and (c) annealed SWCNTs after the irradiation, taken with E_{exc}=1.96 eV(λ_{exc}=632.8 nm). The RBMs were fitted with five Lorentzian lines. One (dotted) line at the lowest frequency is associated with semiconducting SWCNTs. Other (solid) lines at higher frequencies are associated with metallic SWCNTs. [**9**]

In summary, laser-induced defects influence not only D band but also G⁻ peak associated with metallic SWCNTs, which is attributed to the electron-phonon coupling with Kohn anomaly. The upshift and narrowing of the G⁻ peak occur due to the laser irradiation. The G⁻ peak can recover to the original one due to the thermal annealing. The electron-phonon coupling for metallic SWCNTs can be reversibly controlled by the generation and annihilation of specific defects due to the laser irradiation and thermal annealing.

5. FINE STRUCTURE OF D BAND RELATED TO LASER-INDUCED DEFECTS IN COMOCAT SWCNTS

5.1. Heating and Laser Irradiation For Comocat Swcnts

As-received CoMoCAT SWCNTs (SWeNT® CG 100, SouthWest NanoTechnologies, Inc.) were used for heating and laser irradiation experiments. A suspension of SWCNTs in ethanol was prepared by ultrasonication. By drop-coating and air-drying the suspension, a SWCNT thin film was formed on a quartz substrate. For heating experiments, the film samples were annealed at 350 °C for 90 min in air. For laser irradiations, the samples were irradiated with a 532 nm (~2.33 eV) from a $Nd:YVO_4$ laser for 180 min. The irradiation power level in a focal spot of 1 μm in diameter on the sample was kept at ~20 mW. The heating and laser irradiation experiments were also carried out in a vacuum of~4.5 Pa and a dynamic vacuum of ~3.5×10^{-4} Pa, respectively.

5.2. Change In D Band By Heating

Figure 7 shows D and G bands in the Raman spectrum for a pristine CoMoCAT SWCNT sample. The corresponding radial breathing modes (RBMs) are also shown in the inset in the figure. Note that the spectrawere taken with E_{exc}=2.33 eV(λ_{exc}=532 nm) where the laser power level in a focal spot of 1 μm in diameter on the sample was kept below 0.1 mW to prevent overheating the sample.The spectral peaks are fitted with Lorentzian lines. From the RBM frequencies, the mean diameter of pristine SWCNTs is estimated to be ~0.8 nm, which corresponds to typical mean diameter of CoMoCAT ones [27]. Note that the diameter is estimated using ω [cm^{-1}]=234/d [nm]+10 where d is the SWCNT diameter and ω is the RBM frequency [12]. As shown in Fig. 7, the D band in pristine SWCNTs is fitted with two Lorentzian lines at 1313 and 1355 cm^{-1} which are denoted by D_1 and D_3, respectively. The relative intensities of the D components to the most intense G peak at 1594 cm^{-1} in each spectrum are compared in order to clarify the change in the D band.

Table 3. Peak frequencies (ω), full widths at half maximum (Γ), and relative intensities (I/I_{S1}) of Lorentzian lines used to fit radial breathing modes (RBMs) for pristine SWCNTs, laser-irradiated SWCNTs, and annealed SWCNTs after the irradiation in **Fig. 6**. The relative intensities of the peaks are normalized by one located at lowest frequency associated with semiconducting SWCNTs. S and M indicate RBM peaks associated with semiconducting and metallic SWCNTs, respectively.

	pristine SWNTs			laser-irradiated SWNTs			annealed SWNTs		
	ω (cm^{-1})	Γ (cm^{-1})	I/I_{S1}	ω (cm^{-1})	Γ (cm^{-1})	I/I_{S1}	ω (cm^{-1})	Γ (cm^{-1})	I/I_{S1}
S1	157	12	1.0	157	10	1.0	152	9	1.0
M1	168	12	0.52	167	11	0.43	167	19	0.53
M2	173	10	3.2	174	10	0.95	171	12	3.0
M3	186	15	0.59	185	14	0.36	184	14	0.40
M4	196	12	0.42	201	10	0.07	198	14	0.13

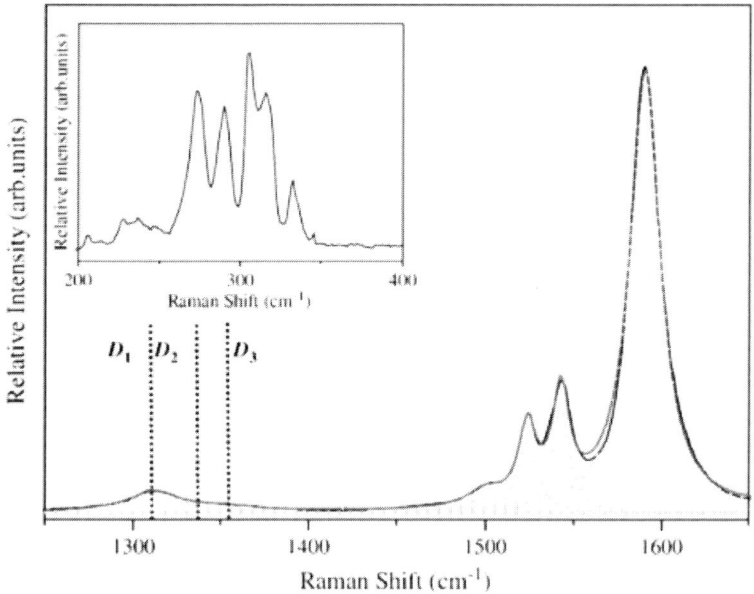

Figure 7. D and G bands in the Raman spectrum for a pristine CoMoCAT SWCNT sample. The corresponding radial breathing modes (RBMs) are also shown in the inset. [**10**].

Figure 8 shows *D* and *G* band in Raman spectra for CoMoCAT SWCNT samples heat-treated at 350 °C for 90 min in (a) air and (b) a vacuum of 4.5 Pa. The corresponding radial breathing modes (RBMs) are also shown in the inset. Note that these spectra were taken with E_{exc}=2.33 eV where the laser power level in a focal spot of 1 μm in diameter on the sample was kept below 0.1 mW to prevent overheating the sample. As shown in Fig. 8(a), the heat-treatment in air leads to the significant increase of the relative intensity of the D_3 component. The corresponding change is also seen in *G*band and RBMs. The broadening of *G* band occurs. This means that the imperfection of SWCNTs increases. Namely, a lot of defects are introduced into SWCNTs. In addition, the higher frequency peaks in the RBMs are disappeared. This means that the degradation of SWCNTs with smaller diameters occurs. On the other hand, no significant change in these Raman bands is observed in SWCNTs heat-treated in a vacuum of ~4.5 Pa as shown in Fig. 8(b). Therefore, it is suggested that the change in the Raman bands by the heat-treatment in air is due to the thermal oxidation.

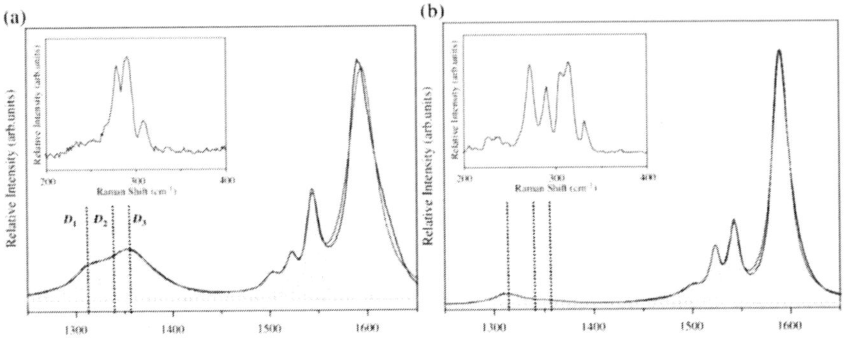

Figure 8. D and G bands in Raman spectra for CoMoCAT SWCNT samples heat-treated at 350 °C for 90 min in (a) air and (b) a vacuum of ~4.5 Pa. The corresponding radial breathing modes (RBMs) are also shown in the inset. [**10**]

Actually, pristine SWCNT sample contains impurities such as amorphous carbon, water, and C–H complex. The thermal oxidation gives rise to open-end structures of SWNTs or holes in the walls [28]. The extension of the oxidation process can generate C=O, C–O–C, and C–OH [29]. In addition, the corresponding *D* band frequency is theoretically predicted to be more than 1353 cm^{-1} [29]. This value is in good agreement with 1355 cm^{-1} of D_3. Thus, the D_3 component can be related to defects such as amorphous carbon and oxides as discussed above.

5.3. Change In D Band By Laser Irradiation

Figure 9 shows *D* and *G* bands in Raman spectra for CoMoCAT SWCNT samples irradiated with E_{laser}=2.33 eV from a Nd:YVO_4 laser with 183 kW/cm^2-for 180 min in (a) air and (b) a dynamic vacuum of~3.5×10^{-4} Pa. The corresponding radial breathing modes (RBMs) are also shown in the inset. Note that these spectra were also taken with E_{laser}=2.33 eV where the laser power level in a focal spot of 1 μm in diameter on the sample was kept below 0.1 mW to prevent overheating the sample. As shown in Fig. 9(a), the laser irradiation in air leads to the increases of the relative intensities of not only D_3 but also D_1. The increase of the D_3 is due to the thermal oxidation, similar to that by the heat-treatment in air as discussed in 5.2, since the laser irradiation occurs the local heating. It should be noted that the D_1 intensity increases by 78%, relative to the pristine one. However, no significant change is seen in G and RBMs. This means that no significant degradation of SWCNTs occurs. These behaviors are quite different from those by the heat-treatment in air as discussed in 5.2. In the laser irradiation in air, H_2 and O_2 radical species can be produced [29,30]. These with water can give rise to C–H complex on the side walls, at defect sites, or ends of SWCNTs [30,31]. Therefore, the D_1 can be related to C–H complex produced by the laser irradiation in air.

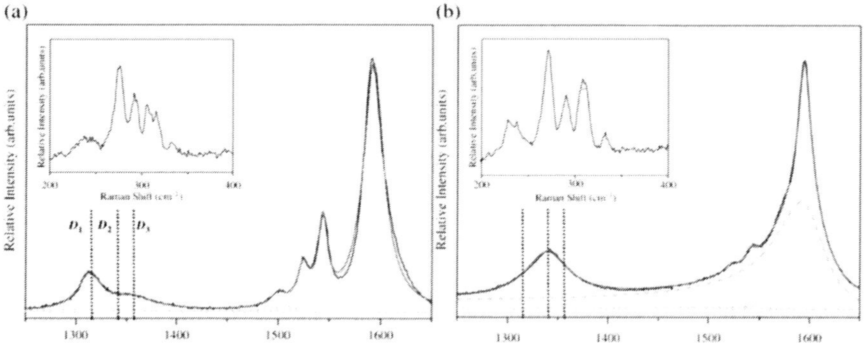

Figure 9. D and G bands in Raman spectra for CoMoCAT SWCNT samples irradiated with E_{laser}=2.33 eV from a Nd:YVO laser for 180 min in (a) air and (b) a dynamic vacuum of~3.5×10^{-4} Pa. The corresponding radial breathing modes (RBMs) are also shown in the inset. [**10**]

On the other hand, the laser irradiation in a dynamic vacuum of ~3.5×10^{-4} Pa leads to the appearance of a new D component at ~1340 cm^{-1}, denoted by D_2, as shown in Fig. 9(b). The peak profile is quite similar to that of step edges in graphite [32]. Therefore, the D_2would be related to the edges of SWCNTs cut by the laser irradiation in the dynamic vacuum. Such cutting might be due to absorbates-assisted burning by laser heating. In addition, it is of interest that the decrease of the D_1 intensity, relative to the pristine one, is observed. The C–H complex formed on SWCNTs is released by thermal annealing at more than 200 °C [30]. The laser irradiation in this experiment gives rise to the local heating. This will lead to the dehydration of SWCNTs so that the D_1 intensity can be decreased. The dehydration is also accompanied by the increases of the intensities of thelower frequency peak at 230 cm^{-1} in the RBMs and of the broad Breigt–Wigner–Fano line in the *G* band, associated with metallic SWCNTs, as seen in Fig. 9(b).

In summary, *D* band in CoMoCAT SWCNTs is composed of three components D_1, D_2, and D_3 at ~1313, 1340, and 1355 cm^{-1}, respectively. These components are attributed to different kinds of defects introduced by heating and laser irradiation. Such insight on the fine structure of the *D* band will play a role for more detailed understanding of *D* band and the identification of defects in SWCNTs.

6. FORMATION OF *TRANS*-POLYACETYLENE FROM COMOCAT SWCNTS BY LASER IRRADIATION

6.1. Laser Irradiation For Comocat Swcnts

As-received CoMoCAT SWCNTs (SWeNT@CG 100, SouthWest NanoTechnologies, Inc.) were used in this experiment. A suspension of SWCNTs in ethanol was prepared by ultrasonication. The suspension was dropped on a clean quartz substrate and allowed to be air-dried at room temperature. The SWCNTs samples prepared in above procedure were used for laser irradiation experiments. The samples, which were exposed to air for less than 1 h before laser irradiation, are called "short air-exposure" ones. Some of samples were kept in air at room temperature for more than six months before laser irradiation. They are called "long air-exposure" ones.

Laser irradiation experiments were carried out using a micro-Raman systemequipped with mirrors, attenuators, a 100× microscope objective, a holographic notch filter, a single grating spectrometer (1800 1/mm grating), and a charge coupled device detector. In the laser irradiation experiments, all attenuators were removed. A 532 nm (~2.33 eV) from a cw Nd:YVO_4 laser was used to irradiate the samples. The laser beam was focused on the sample through the 100× microscope objective, with spot size of 1 μm. The laser power level on the sample was kept at 17.8 mW/μm^2. The irradiation time was 1 h.

6.2. Irradition Effect For "Short Air-Exposure" Comocat Swcnts

Figure 10 shows D and G bands in Raman spectra for a "short air-exposure"CoMoCAT SWCNT sample (a) before and (b) after laser irradiation of a 532 nm with 17.8 mW/μm^2 for 1 h in air. The corresponding RBMs are also shown in the inset in the figure. Note that the spectra excitation was also provided with λ_{exc}=532 nm where the laser power level in a focal spot of 1 μm in diameter on the sample was kept below 0.2 mW/μm^2 to prevent overheating the sample.

The spectral peaks are fitted with Lorentzian lines. From the RBM frequencies in Fig. 10(a), the mean diameter of SWCNTs is estimated to be 0.8 nm, which corresponds to typical mean diameter of CoMoCAT ones [27]. Note that the diameter is estimated using ω [cm^{-1}] = 234/d [nm] + 10 where d is the SWCNT diameterand ω is the RBM frequency [20]. The D band at 1300–1400 cm^{-1} and G band at 1500–1700 cm^{-1} are associated with defects and tangential modes of SWCNTs, respectively. The D band can be fitted with two Lorentzian lines at 1313 and 1346 cm^{-1}. The G band can be fitted with three Lorentzian lines at 1523, 1543, and 1591 cm^{-1}.

After laser irradiation, the increase of relative intensity ofD band is observed as seen in Fig. 10(b). Especially, the lower-frequency component increases. The increase of D band intensity can be attributed to the oxidation and/or hydrogenationof SWCNTs [10,29,30] as discussed in 5. In addition, the RBMs exhibit the decrease of higher-frequency components by laser irradiation. This means that smaller diameter SWCNTs are degraded by thermal oxidation due to laser heating. On the other hand, no significant change in G band is observed as seen in Fig. 10(b).

6.3. Irradiation Effect For "Long Air-Exposure" Comocat Swcnts

Figure 11 shows D and G bands in Raman spectra for a "long air-exposure"CoMoCAT SWCNT sample (a) before and (b) after laser irradiation of a 532 nm with 17.8 mW/μm^2 for 1 h in air. The corresponding RBMs are also shown in the inset in the figure.Note that the spectra excitation was also provided with λ_{exc}=532 nm where the laser power level in a focal spot of 1 μm in diameter on the sample was kept below 0.2 mW/μm^2 to prevent overheating the sample. As seenin Fig. 11(a), the "long air-exposure" sample also exhibits clearD and G bands, and RBMs, fitted by Lorentzian lines, as pristine ones. The peak profiles of G band and RBMs are similar to those of pristine one. On the other hand, the relative intensity of D band increases compared with that of pristine one in Fig. 10(a). The D band intensity is similar to that of the irradiated "short air-exposure" one in Fig. 10(b). This means that the oxidation and/or hydrogenation occur in "long air-exposure" ones even before laser irradiation. This is due to the exposure of air with water vapor for the long time more than six months.

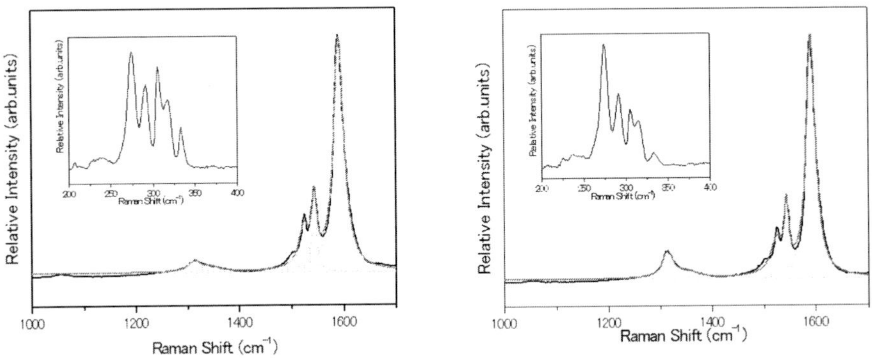

Figure 10. D and G bands in Raman spectra for a "short air-exposure"CoMoCAT SWCNT sample (a) before and (b) after laser irradiation of a 532 nm with 17.8 mW/μm^2 for 1 h in air. The corresponding RBMs are also shown in the inset in the figure. [**11**]

It should be noted that a significant change in Raman spectra is observed for the irradiated "long air-exposure"sample as seen in Fig. 11(b). Namely, new intense peaks appearat 1138 cm^{-1} and 1514 cm^{-1}for the irradiated "long air-exposure" sample. These spectral features are quite similar to those of *trans*-polyacetylene, as have been reported so far [33–36]. According to the model calculation [37], the lower-frequency peak of 1138 cm^{-1} is assigned to a coupled C–C stretching and C–H bending vibration. The higher-frequency peak of 1514 cm^{-1} is assigned to a C–C stretching vibration. In addition, for short polyacetylene chains, the peak position around 1150 cm^{-1} strongly depends on the chain length, and shifts to lower frequency with increasing chainlength [37]. Thus, the appearance of intense peaks of 1138 and 1514 cm^{-1} means that *trans*-polyacetylene-like structural units are formed in irradiated "long air-exposure"samples.

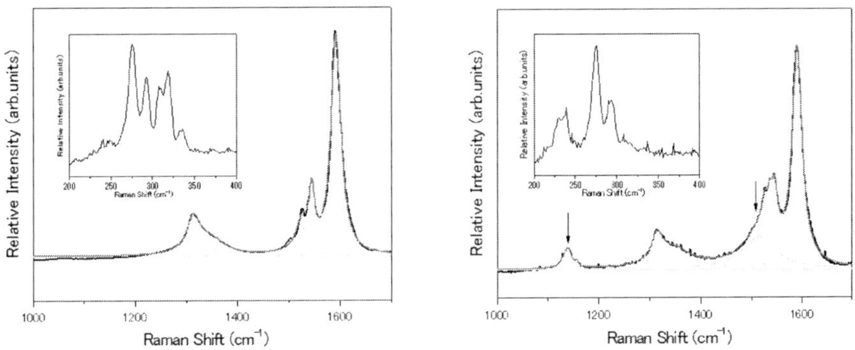

Figure 11. D and G bands in Raman spectra for a "long air-exposure"CoMoCAT SWCNTsample (a) before and (b) after laser irradiation of a 532 nm with 17.8 mW/μm^2 for 1 h in air. The corresponding RBMs are also shown in the inset in the figure. The arrows indicate peaks at 1138 and1514 cm^{-1}associated with*trans*-polyacetylene. [**11**]

6.4. Formation Of Polyacetylene From Swcnts

In "long air-exposure" samples, a lot of water vapor can be absorbed. To remove the absorbed water, the "long air-exposure"samples were annealed at 150 °C for 1 h under a dynamic vacuum of 8.7×10^{-7}Pa. In order to examine the effect of absorbed water on the formation of *trans*-polyacetylene, laser irradiation experiments were also carried out for the annealed "long air-exposure" samples. As a result, for the annealed samples, no clear peak at around 1140 and 1540 cm^{-1} corresponding to *trans*-polyacetylene were appeared even after the laser irradiation. This means that the absorbed water plays a role for the formation of *trans*-polyacetylene.

Let us consider the formation process of *trans*-polyacetylene from "long air-exposure" samples by laser irradiation. It is well-known that CoMoCAT SWCNTs (SWeNT CG 100) include metal catalysts [27]. The catalysts are coated by a carbon layer, and deactivated. The watercan remove the carbon layer at high temperature, and revive the catalytic activity [38]. In "long air-exposure" samples used in this experiment, the water vapor can be sufficiently absorbed on catalysts. The water can remove the carbon layer under laser heating and revive catalytic activity. Consequently, as observed in graphite, graphene and CNT[39-43], the cutting of nanotubes can be realized by catalytic hydrogenation of carbon atoms, in which metal particle dissociate carbon atoms in CNTs, and then the dissociated carbon atoms react with H_2 to create hydrocarbon species such as CH_4. The cutting can be also accompanied with the formation of C–H species on the nanotubes as shown in Figure 12. Such cutting process would lead to the formation of polyacetylene-like structural units.

In summary, *trans*-polyacetylene is formed from CoMoCATSWCNT samples including metal catalysts and absorbed water by laser heating in air. The formation process might be related to the cutting of SWCNTs due to the catalytic hydrogenation of carbon atoms with laser heating, although the detailed mechanism is not yet understood. This shows a new use of the laser irradiation for the formation of functional materials from SWCNTs.

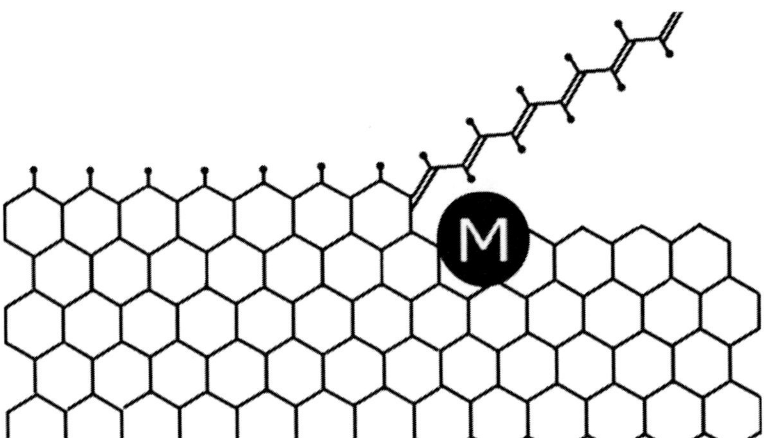

Figure 12. Schematic drawing of the formation of *trans*-polyacetylene by the cutting of SWCNT edge due to the catalytic hydrogenation with laser irradiation.

7. SUMMARY

This chapter presented the characterization of laser-induced defects in SWCNTs by Raman spectroscopy. The laser irradiation with heating followed by burning can produce defects such as vacancies and interstitials in SWCNTs. These defects greatly influence electronic structures and phonon properties especially in metallic nanotubes. They can also be thermally relaxed by vacancy-interstitial recombination and vacancy migration along the tube axis with activation energy of 0.4 eV and 0.7 eV, respectively. This means that the electronic structures and phonon properties in metallic nanotubes can be reversibly controlled by the generation and annihilation of specific defects due to laser irradiation and thermal annealing. In addition, it was presented that the fine structure of Raman D band can be related to specific defects such as C-H complex on nanotubes and nanotube edges produced by laser irradiation. This can lead the Raman spectroscopy to a more effective tool for the characterization of defects in SWCNTs. Finally, it was presented that the laser irradiation can give rise to the formation of trans-polyacetylene from SWCNTs. The formation process might be related to the cutting of SWCNTs due to the catalytic hydrogenation of carbon atoms with laser heating, although the detailed mechanism is not yet understood. This shows a new use of the laser irradiation for the formation of functional materials from SWCNTs. Thus, the laser irradiation is useful for not only the understanding of the properties of defects in SWCNTs but also the modificaton of SWCNTs, as electron and ion irradiations.

ACKNOWLEDGEMENTS

The author acknowledges the contributions to the works presented here by all past and present collaborators, especially Dr. Takashi Uchida, Dr. Hironori Kawamoto, Mr Ken-ichi Kato, Mr. Dongchul Kang, Ms. Mari Hakamatsuka, Ms. Nagisa Hosoya, Mr. Noriaki Nemoto, and emeritus Prof. Kenichi Kojima. The works were supported in part by Strategic Research Projects in Yokohama City University and MEXT/JSPS KAKENHI.

REFERENCES

1. A. V. Krasheninnikov, F. Banhart, 2007 Engineering of nanostructured carbon materials with electron or ion beams Nat. Mater 6 723 733

2. A. V. Krasheninnikov, K. Nordlund, 2010Ion and electron irradiation-induced effects in nanostructured materials J. Appl. Phys 107 071301-1-70

3. S. Suzuki, Y. Kobayashi, 2007 Healing of Low-Energy Irradiation-Induced Defects in Single-Walled Carbon Nanotubes at Room Temperature J. Phys. Chem. C 111 12 4524 4528

4. B. Zandian, R. Kumar, J. Theiss, A. Bushmaker, S. B. Cronin, 2009 Selective destruction of individual single walled carbon nanotubes by laser irradiation Carbon 47 5 1292 1296

5. H. O. Jeschke, M. S. Diakhate, M. E. Garcia, 2009 Molecular dynamics simulations of laser-induced damage of nanostructures and solids Applied Physics A: Materials Science & Processing 96 1 33 42

6. P. Kumar, L.S. Panchakarla, C.N.R. Rao, 2011 Laser-induced unzipping of carbon nanotubes to yield graphenenanoribbons Nanoscale 3 2127 2129

7. M. Mases, M. Noël, M. Dossot, E. Mc Rae, V. Alexander, 2011 Laser-induced damage and destruction of HiPCO nanotubes in different gas environments physica status solidi 248 11 2540 2543

8. T. Uchida, M. Tachibana, K. Kojima, 2007 Thermal relaxation kinetics of defects in single-wall carbon nanotubes J. Appl. Phys 101 084313-1-4

9. D. Kang, K. Kato, K. Kojima, T. Uchida, 2008 Tachibana M. Phonon control in metallic carbon nanotubes due to laser-induced defects Appl. Phys. Lett 93 133102-1-3

10. D. Kang, M. Hakamatsuka, K. Kojima, M. Tachibana, 2010 Influence of heating and laser irradiation on the Raman D band in single-wall carbon nanotubes Diamond & Related Materials 19 578 580

11. M. Hakamatsuka, H. Yoshimura, M. Tachibana, 2011 Formation of trans-polyacetylene from single-wall carbon nanotubes Carbon 46 1869 1872

12. A. Jorio, M. Dresselhaus, R. Saito, G. F. Dresselhaous, 2011 Raman Spectroscopy in Graphene Related Systems Weinheim: Wiley-VCH

13. S. Bandow, S. Asaka, Y. Saito, A. M. Rao, L. Grigorian, E. Richte, P. C. Eklund, 1998 Effect of the Growth Temperature on the Diameter Distribution and Chirality of Single-Wall Carbon Nanotubes Phys. Rev. Lett 80 3779 3782

14. D. M. Brown, A. Jorio, P. Corio, M. S. Dresselhaus, G. Dresselhaus, R. Saito, K. Kneipp, 2001 Origin of the Breit-Wigner-Fanolineshape of the tangential G-band feature of metallic carbon nanotubes Phys. Rev. B 63 155414 155421

15. V. H. Crespi, N. G. Chopra, M. L. Cohen, A. Zettl, S. G. Louie, 1996 Anisotropic electron-beam damage and the collapse of carbon nanotubes Phys.Rev. B 54 5927 5931

16. E. Asari, M. Kitajima, K. G. Nakamura, T. Kawabe, 1993 Thermal relaxation of ion-irradiation damage in graphite Phys. Rev. B 47 11143 11148

17. A. V. Krasheninnikov, P. O. Lehtinen, A. S. Foster, R. M. Nieminen, 2006 Bending the rules: Contrasting vacancy energetics and migration in graphite and carbon nanotubes Chem. Phys. Lett 418 132 136

18. M. Lazzeri, S. Piscanec, F. Mauri, A. C. Ferrari, J. Robertson, 2006 Phonon linewidths and electron-phonon coupling in graphite and nanotubes J. Phys. Rev. B 73 155426 155431

19. S. Piscanec, M. Lazzeri, J. Robertson, A. C. Ferrari, F. Mauri, 2007 Optical phonons in carbon nanotubes: Kohn anomalies, Peierls distortions, and dynamic effects Phys.Rev. B 75 035427 035448

20. M. Milnera, J. Kurti, M. Hulman, H. Kuzmany, 2000 Periodic Resonance Excitation and Intertube Interaction from Quasicontinuous Distributed Helicities in Single-Wall Carbon Nanotubes Phys. Rev. Lett 84 1324 1327

21. H. Kataura, Y. Kumazawa, Y. Maniwa, I. Umezu, S. Suzuki, 1999 Ohtsuka Y.,Achiba Y. Optical properties of single-wall carbon nanotubes. Synth Met 103 2555 2558

22. A. Jorio, C. Fantini, M. A. Pimenta, R. B. Capaz, G. G. Samsonidze, G. Dresselhaus, M. S. Dresselhaus, J. Jiang, N. Kobayashi, A. Gruneis, R. Saito, 2005 Resonance Raman spectroscopy (n,m)-dependent effects in small-diameter single-wall carbon nanotubes Phys. Rev. B 71 075401 075411

23. Y. Li, S. V. Rotkin, U. Ravaioli, 2004 Metal-semiconductor transition in armchair carbon nanotubes by symmetry breaking Appl. Phys. Lett 85 4178 4180

24. Y. , P. O. Lehtinen, A. S. Foster, R. M. Nieminen, 2004 Magnetic properties of vacancies in graphene and single-walled carbon nanotubes New J. Phys 6 68

25. A. Vijayaraghavan, K. Kanzaki, S. Suzuki, Y. Kobayashi, H. Inokawa, Y. Ono, S. Kar, P. M. Ajayan, 2005 Metal–Semiconductor Transition in Single-Walled Carbon Nanotubes Induced by Low-Energy Electron Irradiation Nano Lett 5 1575 1579

26. Park J.Y. 2007 Electrically tunable defects in metallic single-walled carbon nanotubes Appl. Phys. Lett 90 2 023112 023114

27. S. M. Bachilo, L. Balzano, J. E. Herrera, F. Pompeo, D. E. Resasco, R. B. Weisman, 2003 Narrow (n,m)-Distribution of Single-Walled Carbon Nanotubes Grown Using a Solid Supported Catalyst J. Am. Chem. Soc. 125 37 11186 11187

28. P. X. Hou, C. Liu, H. M. Cheng, 2008 Purification of carbon nanotubes Carbon 46 15 2003 2025

29. A. M. D. Silva, G. M. A. Junqueira, C. P. A. Anconi, H. F. D. Santos, 2009 New Insights on Chemical Oxidation of Single-Wall Carbon Nanotubes: A Theoretical Study Phys. Chem. C 113 23 10079 10084

30. G. Zhang, P. Qi, X. Wang, Y. Lu, D. Mann, X. Li, H. Dai, 2006 Hydrogenation and Hydrocarbonation and Etching of Single-Walled Carbon Nanotubes J. Am. Chem. Soc. 128 18 6026 6027

31. A. Nikitin, H. Ogasawara, D. Mann, R. Denecke, Z. Zhang, H. Dai, K. Cho, A. Nilsson, 2005 Hydrogenation of Single-Walled Carbon Nanotubes Phys. Rev. Lett. 95 22 225507 225510

32. L. G. Cancado, M. A. Pimenta, B. R. A. Neves, M. S. S. Dantas, A. Jorio, 2004 Influence of the Atomic Structure on the Raman Spectra of Graphite Edges Phys. Rev. Lett. 93 24 247401 247404

33. H. Kuzmany, P. Knoll, 1986 Resonance Raman cross-section and polarization of light scattered from trans-polyacetylene Raman Spectroscopy 17 1 89 92

34. H. Takeuchi, T. Arakawa, Y. Furukawa, I. Harada, H. Shirakawa, 1987 Density of vibrational states in trans-polyene: comparison with the infrared, Raman and neutron spectra of trans-polyacetylene J MolStruct 158 179 193

35. A. C. Ferrari, J. Robertson, 2001 Origin of the 1150-cm 1 Raman mode in nanocrystalline diamond Phys Rev B 63 12 121405 121408

36. H. Kuzmany, R. Pfeiffer, N. Salk, B. Günther, 2004 The mystery of the 1140 cm 1 Raman line in nanocrystalline diamond films Carbon 42 5-6 911 917.

37. Owens F.J. 2005 Effect of nanosizing on some properties of one dimensional polyacetylene chains Physica E 25 4 404 408

38. T. Yamada, A. Maigne, M. Yudasaka, K. Mizuno, D. N. Futaba, M. Yumura, S. Iijima, K. Hata, 2008 Revealing the secret of water-assisted carbon nanotube synthesis by microscopic observation of the interaction of water on the catalysts Nano Lett 8 12 4288 4292

39. A. Tomita, Y. Tamai, 1974 An optical microscopic study on the catalytic hydrogenation of graphite J PhysChem 78 22 2254 2258

40. S. S. Datta, D. R. Strachan, S. M. Khamis, A. T. C. Johnson, 2008 Crystallographic etching of few-layer graphene Nano Lett. 8 7 1912 1915

41. L. C. Campos, V. R. Manfrinato, J. D. Sanchez-Yamagishi, J. Kong, P. Jarillo-Herrero, 2009 Anisotropic etching and nanoribbon formation in single-layer graphene Nano Lett. 9 7 2600 2604

42. L. Ci, L. Song, D. Jariwala, A. L. Elías, W. Gao, M. Terrones, P. M. Ajayan, 2009 Graphene shape control by multistage cutting and transfer Adv Mater. 21 41 4487 4491

43. A. L. Elías, A. R. Botello-Méndez, D. Meneses-Rodríguez, V. J. González, D. Ramírez-González, L. Ci, E. Muñoz-Sandoval, P. M. Ajayan, H. Terrones, M. Terrones, 2010 Longitudinal cutting of pure and doped carbon nanotubes to form graphitic nanoribbons using metal clusters as nanoscalpels Nano Lett. 10 2 366 372

4

GROWTH OF SEMICONDUCTING SINGLE-WALL CARBON NANOTUBES WITH A NARROW BAND-GAP DISTRIBUTION

Feng Zhang[1], Peng-Xiang Hou[1], Chang Liu[1], Bing-Wei Wang[1], Hua Jiang[2], Mao-Lin Chen[1], Dong-Ming Sun[1], Jin-Cheng Li[1], Hong-Tao Cong[1], Esko I. Kauppinen[2] & Hui-Ming Cheng[1,3]

1 Shenyang National Laboratory for Materials Science, Advanced Carbon Division, Institute of Metal Research, Chinese Academy of Sciences, 72 Wenhua Road, Shenyang 110016, China.

2 Nano Materials Group, Department of Applied Physics and Center for New Materials, School of Science, Aalto University, PO Box 15100, FI-00076 Aalto, Finland.

3 Faculty of Science, Chemistry Department, King Abdulaziz University, Jeddah 21589, Saudi Arabia

The growth of high-quality semiconducting single-wall carbon nanotubes with a narrow band-gap distribution is crucial for the fabrication of high-performance electronic devices. However, the single-wall carbon nanotubes grown from traditional metal catalysts usually have diversified structures and properties. Here we design and prepare an acorn-like, partially carbon-coated cobalt nanoparticle catalyst with a uniform size and structure by the thermal reduction of a$[Co(CN)_6]^{3-}$ precursor adsorbed on a self-assembled block copolymer nanodomain. The inner cobalt nanoparticle functions as active catalytic phase for carbon nanotube growth, whereas the outer carbon layer prevents the aggregation of cobalt nanoparticles and ensures a perpendicular growth mode. The grown single-wall carbon nanotubes have a very narrow diameter distribution centred at 1.7 nm and a high semiconducting content of >95%. These semiconducting single-wall carbon nanotubes have a very small band-gap difference of ~0.08 eV and show excellent thin-film transistor performance.

1. INTRODUCTION

Single-wall carbon nanotubes (SWCNTs) have a small diameter, high carrier mobility, tunable band-gap and good stability[1, 2, 3]. Therefore, they are considered to be an ideal channel material for high-performance field effect transistors (FETs), integrated circuits and related electronic devices[4, 5, 6]. However, slight changes in the diameter and chiral angle of a SWCNT lead to the change of its type of electrical conductivity, that is, semiconducting to metallic or *vice versa*. As a result, as-prepared SWCNTs are usually a mixture of semiconducting and metallic nanotubes. In recent times, great efforts have been devoted to the selective preparation of pure semiconducting SWCNTs (s-SWCNTs) in large scale for their potential use in new-generation integrated circuits and notable progress has been made. One way is to separate SWCNTs by postsynthesis chemical and physical treatments[7, 8, 9, 10]. Although high-purity s-SWCNTs can be obtained, this approach suffers from drawbacks that contamination and defects are inevitably introduced and the resultant SWCNTs are usually shortened. The other way is to synthesize s-SWCNTs directly by using metal nanoalloy as catalyst[11, 12] or *in situ* selective etching during growth based on the principle that metallic SWCNTs (m-SWCNTs) are chemically more active than s-SWCNTs[13, 14]. For example, Zhang and colleagues[15] grew s-SWCNT arrays by applying ultraviolet irradiation and subsequent photochemical reactions; Liu and colleagues[16] prepared SWCNTs by using a mixture of ethanol and methanol as both carbon source and oxidant precursor; Cheng and colleagues[17, 18] synthesized s-SWCNTs by introducing oxygen or hydrogen as etchants; and Li and colleagues[19] grew s-SWCNTs from a metal catalyst loaded on CeO_2supports that can release oxygen for selective etching. In these above studies, high-quality s-SWCNTs with purities higher than 90% were obtained. However, little attention has been paid to the band-gap uniformity of the s-SWCNTs obtained. In fact, the diameters of these s-SWCNTs are usually distributed over a wide range. According to theoretical and experimental studies, the band-gap of s-SWCNT is inversely proportional to their diameter[2]. It is only by using s-SWCNTs with a narrow band-gap range that devices with a uniform and stable performance can be fabricated. From this point of view, it is most desirable to grow SWCNTs with uniform chirality. Progress has been made on the controlled growth of SWCNTs by catalyst engineering[20, 21, 22]. Harutyunyan *et al.*[20] first achieved preferential growth of m-SWCNTs by modifying the morphology and coarsening behaviour of catalyst nanoparticles. Scott *et al.*[23] synthesized short (5, 5) CNTs from a corannulene seed by stepwise chemical method and by Diels–Alder reaction[24]. Theoretically, Ding *et al.*[25] proposed that the growth rate of an SWCNT is proportional to its chiral angle, whereas Artyukhov *et al.*[26] pointed out that the naturally enriched near-armchair SWCNTs can be attributed to a 'compromise' of the kinetic and thermodynamic growth trends for a rigid state catalyst. Very recently, high-purity single-chirality SWCNTs were grown from high-melting-point, W-containing alloy nanoparticles[21, 22], as well as growth seeds derived from molecular precursors[27]. However, the enriched (12, 6)[21], (6, 6)[27] and so on SWCNTs are metallic, whereas the content of (16, 0) s-SWCNTs is ~80% (ref. 22).

To obtain s-SWCNTs with a narrow diameter distribution and a small range of band-gap, it is essential to use uniform catalyst nanoparticles. However, it was found that even with a similar initial catalyst nanoparticle size, the grown SWCNTs may have different diameters[28, 29]. There are mainly two reasons for this phenomenon: the first is that catalyst nanoparticles tend to aggregate, to form larger ones during the growth of SWCNTs at high temperatures; second, SWCNTs may grow following either the 'tangential' mode (with diameter equal to the nanoparticle size)[28] or the 'perpendicular' mode (with diameter smaller than the nanoparticle size)[28]. Therefore, to achieve well-controlled growth of SWCNTs, it is crucial to prevent the aggregation of catalyst nanoparticles and to control their mode of nucleation and growth. In this study, we design and prepare an acorn-like bicomponent catalyst, that is, Co nanoparticle with a partial carbon coating layer. The inner Co nanoparticle functions as active catalytic phase for SWCNT growth, whereas the outer carbon layer prevents the aggregation of Co nanoparticles and ensures the perpendicular growth of SWCNT from the catalyst. This acorn-like catalyst with uniform size and structure is prepared using a block copolymer (BCP) self-assembly technique[30]. The SWCNTs grow from the catalyst have a very narrow diameter distribution. Based on the principle that m-SWCNTs are chemically more reactive than s-SWCNTs with similar diameters due to the smaller ionization potential of the former[31, 32], hydrogen is introduced as an etchant and m-SWCNTs are selectively removed *in situ*. As a result, high-purity (>95%) s-SWCNTs with a narrow range of band gaps (<0.08 eV) are obtained, which show an excellent thin-film transistor (TFT) performance.

2. RESULTS

2.1 Preparation of Co nanoparticle partially coated with carbon

A schematic showing the preparation of the Co nanoparticles partially coated with carbon is shown in Fig. 1. Details of catalyst preparation are described in Methods. In brief, an asymmetric poly-(styrene-block-4-vinylpyridine) (PS-b-P4VP) film (Fig. 1a) was self-assembled into vertical P4VP nanocylinders by placing it in a toluene/tetrahydrofuran (THF) mixture vapor annealing for 24 h. The formed nanodomains, consisting of hydrophilic vertical P4VP nanocylinders enclosed by a hydrophobic PS matrix, were then immersed in an acidic aqueous solution of $K_3[Co(CN)_6]$(Fig. 1b). The protonated pyridinic nitrogen sites in the P4VP nanocylinders attracted anionic Co complexes as $[Co(CN)_6]^{3-}$, after which the material was treated in an air plasma to oxidize the polymer. As the Co precursor was anchored at one end of the nanodomains, CoO nanoclusters partially encapsulated by a thin polymer layer were obtained (Fig. 1c). It is worth noting that we intentionally used a relatively weak plasma treatment to remove the polymer in a controlled manner. The PS-b-P4VP used, PS40000-b-P4VP5600, has a high PS to P4VP mass ratio (~7), which also makes it difficult to completely remove the polymer. Furthermore, owing to the similar size of the self-assembled nanodomains, the resulting CoO clusters have a uniform size and exposed surface area. In subsequent heating in a H_2 atmosphere, the CoO nanoclusters were reduced to Co and aggregated into Co nanoparticles, and the residual polymer was carbonized to produce Co catalysts that were partially coated with carbon; thus, the acorn-like bicomponent catalyst is obtained (Fig. 1d). SWCNTs with a uniform diameter (Fig. 1e) were grown from these catalyst particles by the chemical vapour deposition (CVD) of ethanol. When H_2 was introduced as an *in situ* growth etchant, m-SWCNTs with higher chemical reactivity were selectively removed and s-SWCNTs with a narrow range of band-gap were obtained (Fig. 1f).

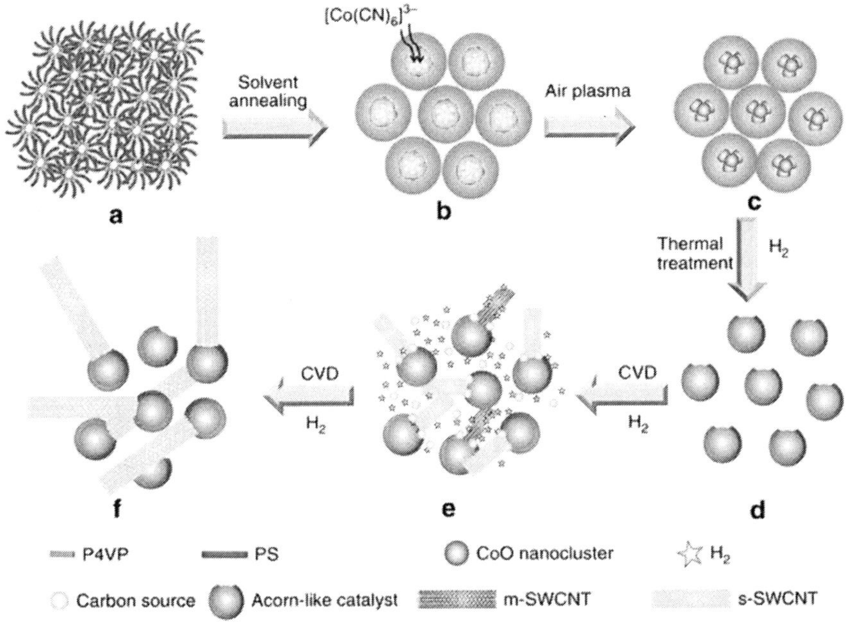

Figure 1. Schematic showing the formation of carbon-coated Co nanoparticles and the growth of s-SWCNTs. (**a**) A self-assembled PS-b-P4VP film. (**b**) Formation of phase-separated nanodomains and adsorption of $[Co(CN)_6]^{3-}$ catalyst precursors. (**c**) CoO nanoclusters partially surrounded by residual polymer. (**d**) Co nanoparticles partially coated with carbon. (**e**) Nucleation and growth of SWCNTs with a narrow diameter distribution grown from the partially carbon-coated Co nanoparticles and *in situ* etching of m-SWCNTs. (**f**) s-SWCNTs with a narrow band-gap range.

2.2 Characterization of the partially coated Co nanoparticles

To understand the structure of the catalyst synthesized using this method, atomic force microscopy (AFM) and transmission electron microscopy (TEM) characterization were performed. An AFM image (Fig. 2a) shows that the catalyst particles are well-dispersed and uniformly distributed on the substrate, whereas the TEM image (Fig. 2b) verifies the uniform size of the monodispersed Co nanoparticles. These nanoparticles were observed to be partially coated with a carbon layer (indicated by red arrows in Fig. 2b). High-resolution TEM images

(Supplementary Fig. 1a–c) clearly demonstrate the partially carbon-coated Co nanoparticle structure of the catalyst. More detailed characterizations of the catalysts are reported in Methods. The sizes of ~130 nanoparticles were measured from TEM images and the resultant histogram of their diameters is shown in Fig. 2c. It can be seen that most of the nanoparticles (>90%) have diameters in the range of 2.5–4.5 nm, with a mean diameter of 3.1 nm. These uniform and monodispersed Co catalyst particles partially coated with carbon are critical for the growth of SWCNTs with a narrow diameter distribution.

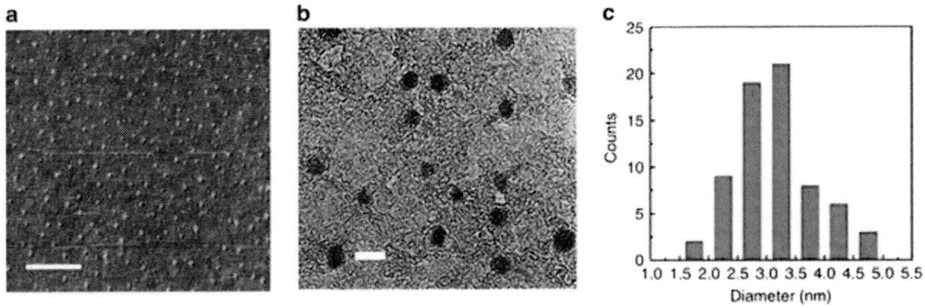

Figure 2. Morphology and diameter distribution of the Co nanoparticle partially coated with carbon**.** Typical (**a**) AFM and (**b**) TEM images of the Co nanoparticles partially coated with carbon, showing that the nanoparticles are monodispersed and uniform in size. The red arrows indicate Co nanoparticles partially coated by a carbon layer. Scale bar, 400 nm (**a**) and 5 nm (**b**). (**c**) A histogram showing the diameter distribution of the nanoparticles based on TEM observations.

2.3 Structure characterization of SWCNTs

SWCNTs were synthesized by CVD at 700 °C using the carbon-coated Co particles as catalyst, alcohol as carbon source and hydrogen as carrier gas. Details of the SWCNT synthesis procedure and characterization are described in Methods. Scanning electron microscope (SEM) image of Fig. 3a shows that carbon nanotubes with a length of 10 μm are randomly dispersed on the substrate. TEM observations show that the SWCNTs obtained are isolated and straight (Fig. 3b), suggesting that the catalyst has a good catalytic activity for SWCNT growth even at the relatively low temperature of 700 °C. A SWCNT grown from a partially carbon-coated catalyst nanoparticle was clearly observed (Fig. 3c). The diameters of 130 SWCNTs were measured from TEM images and the resultant histogram is shown in Fig. 3d. The diameters are narrowly distributed in the range of 1.6–1.9 nm with a mean diameter of 1.7 nm. This diameter distribution is much narrower than most previously reported results[17, 33]. In addition, the SWCNTs are isolated rather than in bundles, which can be attributed to the monodispersed Co catalysts used.

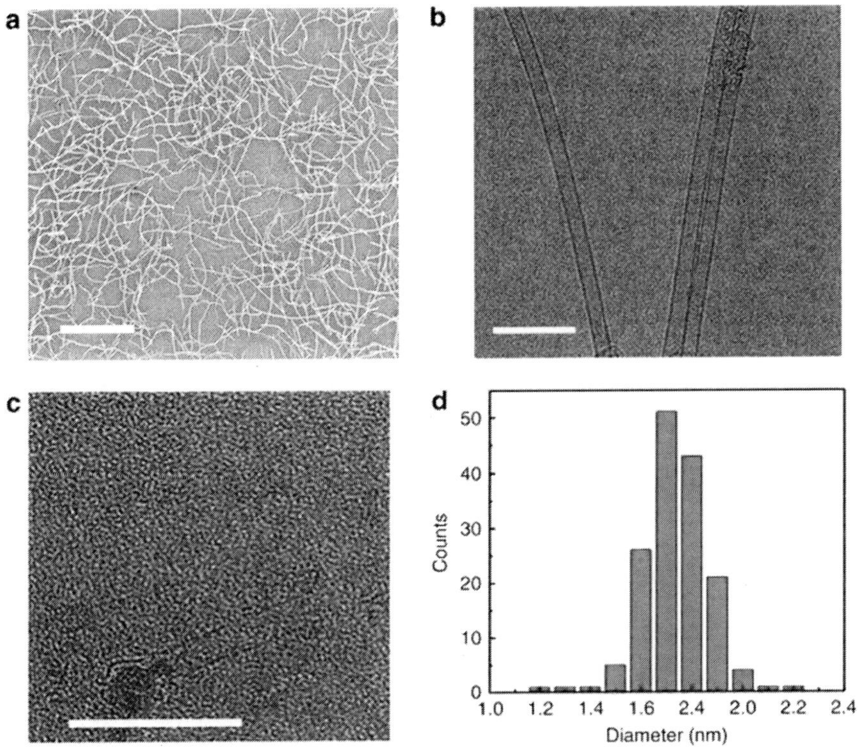

Figure 3. SEM and TEM observations of the as-prepared SWCNTs. SEM (**a**) and TEM (**b**,**c**) image of the as-prepared SWCNTs. Scale bar, 10 μm (**a**) and 10 nm (**b**,**c**). (**d**) Diameter distribution of the SWCNTs based on TEM observations.

2.4 Characterization by Raman and absorption spectroscopy

We further characterized the SWCNTs using Raman spectroscopy with excitation laser wavelengths of 488, 532, 633 and 785 nm. Figure 4a–d show the Raman spectra, where the radial breathing mode (RBM) peaks originating from m- and s-SWCNTs are highlighted according to the Kataura plot[34]. In Fig. 4a, sharp peaks narrowly centred at ~141 cm^{-1} are shown with an excitation wavelength of 532 nm (2.33 eV) and these peaks originate from s-SWCNTs. Figure 4bshows the Raman spectra excited with a laser wavelength of 633 nm (1.96 eV) and the RBM peaks originating from s-SWCNTs are located in a narrow range around ~141 cm^{-1}. Raman spectra excited by a 785 nm (1.58 eV) laser (Fig. 4c) shows very few weak RBM peaks at ~141 cm^{-1} originating from metallic tubes. No RBM Raman peak is detected with the 488-nm laser (2.54 eV) (Fig. 4d). All the excited Raman RBM signals under these four laser wavelengths are centred at ~141 cm^{-1} and most can be assigned to s-SWCNTs, suggesting that the sample contains s-SWCNTs with a narrow diameter distribution. Using the inverse relationship between RBM peak frequency (ω_{RBM}) and tube diameter (d_t) (d_t=235.9/(ω_{RBM}−5.5))[35], the calculated diameters of the SWCNTs are around 1.7 nm, which is in good agreement with the TEM observations. The content of s-SWCNTs in our sample is calculated to be ~98% based on the number of the metallic and semiconducting RBM peaks averaged from 160 Raman spectra[19, 36]. The G-band of the Raman spectra (Fig. 4e) show an obvious Lorentzian line shape, which is characteristic of s-SWCNTs[37] and this further verifies the effective enrichment of semiconducting nanotubes. As the diameters of the SWCNTs are mostly in the range of 1.6–1.9 nm, these s-SWCNTs have an extremely small band-gap difference of 0.08 eV (ref. 38). The ultraviolet/visible/near-infrared absorption spectrum of an aqueous SWCNT dispersion collected from 125 SWCNT samples (for more details, see Methods and Supplementary Fig. 2) is shown inFig. 4f. One intense narrow peak and two weak peaks distributed in the range of 1,200–1,430 nm and 600–715 nm, respectively, were detected, corresponding to the second and third van Hove singularity transition of s-SWCNTs (S_{22} and S_{33}) with diameter distributions of 1.6–1.9 nm (refs34, 38). No obvious first van Hove singularity transition of m-SWCNTs (800–950 nm) was detected by the ultraviolet/visible/near-infrared, confirming the high purity of s-SWCNTs in our sample. Quantitatively, the content of s-SWCNTs in the sample was calculated to be ~ 99% (ref.39).

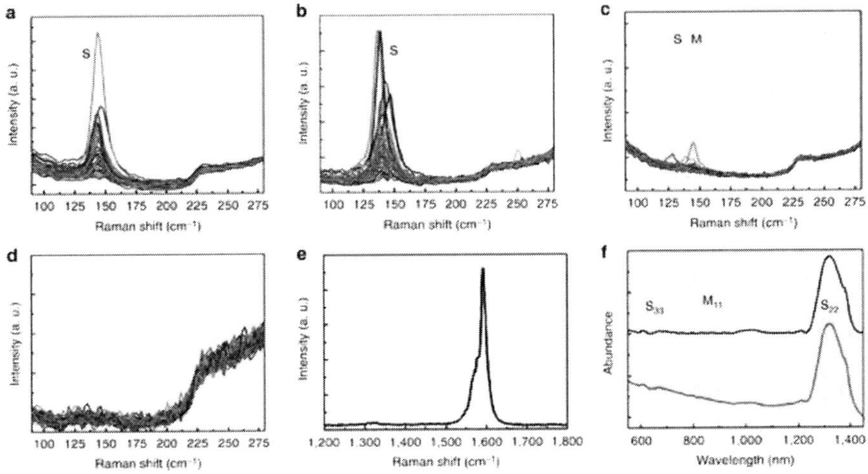

Figure 4. Electrical type of the SWCNTs. Electrical type characterized by multi-wavelength Raman spectroscopy and ultraviolet/visible/near-infrared absorption spectroscopy. Raman spectra of the SWCNTs measured with (**a**) 532, (**b**) 633, (**c**) 785 and (**d**) 488 nm lasers. The regions corresponding to semiconducting and metallic transitions are labelled as S (pink) and M (blue), respectively. (**e**) D and G band excited with a 633-nm laser. (**f**) Ultraviolet/visible/near-infrared absorption spectrum of an aqueous dispersion collected from 125 SWCNT samples: as-collected curve (red line) and background-subtracted curve (black line).

2.5 Conductivity characterization by electron diffraction

Electron diffraction analysis is an effective method for directly determining the chiral indices (n,m) of SWCNTs[40]. To further confirm the enrichment of s-SWCNTs in our sample, we transferred the SWCNTs from a Si substrate to a TEM grid (for details see Methods) and carried out electron diffraction using a TEM (JEOL JEM-2200FS 2 × CS corrected) operated at 80 kV with a Gatan 794 multiscan charge-coupled device camera (1 k × 1 k). Figure 5a shows a freestanding SWCNT and its corresponding electron diffraction pattern is shown in Fig. 5b. In addition to the bright spot at the centre caused by the direct electron beam, the diffraction pattern mainly consists of a set of parallel diffracted layer lines, which are separated by certain distances from the equatorial layer line at the centre. By analysing the layer line distance, the chiral index of this SWCNT was determined to be (19, 5) with a tube diameter of 1.72 nm. The chiral indexes of total 95 SWCNTs were assigned by using this electron diffraction method (Supplementary Figs 3 and 4). The resultant numbers of s- and m-SWCNTs are shown in Fig. 5c. The s-SWCNT content reaches ~96%, which verifies the enrichment of s-SWCNTs in our sample.

Figure 5. Electron diffraction of the SWCNTs. (**a**) Typical TEM image of an isolated SWCNT. Scale bar, 10 nm. (**b**) Electron diffraction pattern of the SWCNT shown in **a**, the chirality of which is assigned as (19, 5). (**c**) A pie chart showing the content of s- and m-SWCNTs based on electron diffraction measurements of a total of 95 SWCNTs.

2.6 Electrical performance of SWCNT TFTs

Electrical measurements can provide direct evidence of the transport properties of SWCNTs[41,42]. Thus, TFT devices based on the SWCNTs with a growth time of 25 min were fabricated (for more details, see Methods). Supplementary Fig. 5 shows a typical SEM image of a long channel bottom-gated SWCNT TFT. It can be seen that the SWCNTs are pure and randomly distributed. The channel width and length of the TFTs are 100 and 200 μm, respectively. Figure 6a shows typical output characteristics (I_{ds}–V_{ds})of the TFTs, which indicates the formation of ohmic contacts between SWCNTs and metal eletrode. Figure 6b shows transfer characteristics (I_{ds}–V_{gs}) of ten devices with the same channel geometry (L_{ch}=200 μm, W_{ch}=100 μm); simultaneously, high on/off ratios of 3.1×10^3–3.6×10^6 and high carrier mobilities of 36–143 $cm^2 V^{-1} s^{-1}$, as evaluated by the standard formula[6], were demonstrated. Remarkably, the mean carrier mobility is as high as 95.2 $cm^2 V^{-1} s^{-1}$. Furthermore, as shown in Fig. 6c, when compared with the previously reported SWCNT-based TFTs[6, 8, 18, 43, 44, 45, 46, 47, 48, 49], including the TFTs fabricated using chirality-enriched nanotubes[46], our TFTs show obviously superior performance, in terms of both high current on/off ratio and high carrier mobility, which further verifies the high quality and high s-SWCNTs content of our sample. We further constructed a series of bottom-gate TFTs from the grown s-SWCNTs with L_{ch} ranging from 20 to 200 μm (Supplementary Fig. 6). It was found that on/off ratios higher than 10^5 were obtained for the TFTs with L_{ch} longer than 40 μm, and even for the devices with a L_{ch} value of 20 μm, the on/off ratios were still around 10^2. TFTs with a short channel length of 1.5 μm were also fabricated by electron-beam lithography (EBL) using the same sample. High on/off ratios and high carrier mobilities were demonstrated for the majority of the devices (Supplementary Fig. 7), although a few TFTs showing current on/off ratios <5 caused by the short circuiting of m-SWCNTs were also detected.

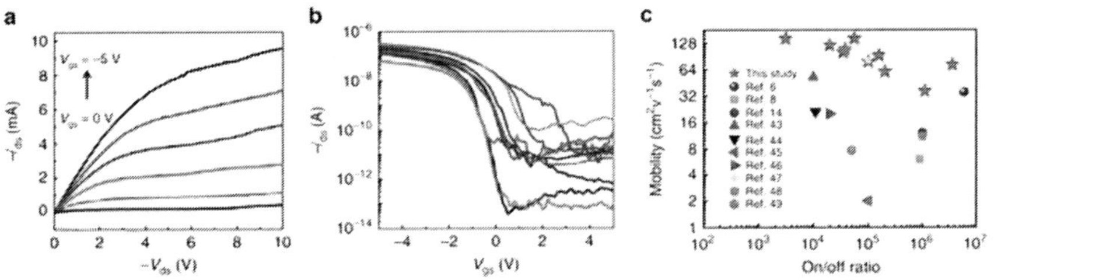

Figure 6. Electrical performance of the SWCNT TFTs. (**a**) Output (I_{ds}−V_{ds}) characteristics of the same device measured at various V_{gs} from −5 to 0 V in 1 V step. L_{ch}=200 μm, W_{ch}=100 μm. (**b**) Transfer characteristics of 10 SWCNT TFTs. (**c**) A comparison of the mobility performance of our SWCNT TFTs with those previously reported.

We attribute this selective growth of s-SWCNTs to both novel catalyst and *in situ* growth selective etching. As mentioned above, we prepared an acorn-like, partially carbon-coated Co nanoparticle catalyst with a narrow diameter distribution by a BCP self-assembly method. The$[Co(CN)_6]^{3-}$ catalyst precursor was absorbed on the phase-separated P4VP nanodomains so that only part of the Co nanoparticle is exposed and serves as the catalyst for SWCNT growth following a perpendicular mode (see Fig. 1). As shown in Fig. 3c, SWCNTs were observed to grow from the partially carbon-coated Co nanoparticles via a perpendicular growth mode. As the Co nanoparticles prepared by the copolymer self-assembly method have a narrow range of diameters, the exposed Co active catalyst obtained after polymer wrapping and partial etching processes should have an even smaller size difference. As a consequence, the SWCNTs grown on the exposed Co regions have an extremely narrow diameter distribution. The hydrogen used as carrier gas during the CVD growth also serves as an etchant, that is, m-SWCNTs can be preferentially removed by H radicals decomposed from H_2 at 700 °C in the presence of Co[50]. The narrow diameter distribution of the SWCNTs guarantees efficient selective etching of m-SWCNTs, as the chemical stability of SWCNTs is determined by both their electrical type and diameter[14,16, 17, 31, 32, 51, 52]. Therefore, high purity (>95%) s-SWCNTs with a uniform diameter and band-gap are obtained.

To verify the above controlled growth mechanism, we further performed a series of comparative experiments. First, fully exposed Co catalyst particles were prepared by omitting the solvent annealing process and a thermal treatment at 750 °C in air for 5 min was added to completely remove the polymer (see Methods). These Co catalysts were used for growing SWCNTs. The Raman spectra (Supplementary Fig. 8) of the obtained sample show a broad RBM peak distribution, originating from both s-and m-SWCNTs according to the Kataura plot[34]. This result proves that the partial carbon coating of the Co nanoparticles plays a key role in growing SWCNTs with uniform diameters. Following this controlled growth mechanism, it may be feasible to tune the diameter of SWCNTs by controlling the exposed area of the Co nanoparticles. We therefore tried to enlarge the exposed area by performing heat treatment in a H_2 atmosphere at 800 °C for 10 min and then used the resulting material for SWCNT growth (Methods).Supplementary Fig. 9 shows the Raman spectra of the SWCNTs obtained. It can be seen that the RBM peaks are distributed in the range of 110–120 cm^{-1}, corresponding to nanotube diameters of 2.0–2.2 nm. Based on the Raman spectra from three laser wavelengths, the content of s-SWCNT was calculated to be ~96% (refs 19, 34, 36). This result further verifies the growth mechanism of s-SWCNTs from the Co catalyst partially coated with carbon.

In summary, acorn-like partially carbon-coated Co nanoparticles with a mean diameter of 3.1 nm were prepared by a BCP self-assembly and subsequent partial oxidation approach. The bicomponent nanoparticles were used as a catalyst for growing SWCNTs by CVD. SWCNTs with a narrow and tunable diameter distribution were synthesized in a hydrogen atmosphere following a perpendicular growth mode. It was found that a high content (>95%) of s-SWCNTs with a narrow diameter distribution was obtained. The mean diameter of the s-SWCNTs could be tuned to be ~1.7 or 2.1 nm by controlling the carbon coverage of the Co catalyst. The band-gap range of the s-SWCNTs was <0.08 eV, due to their very narrow diameter distribution. The TFTs fabricated using these s-SWCNTs showed an excellent electrical performance. This combined catalyst design and *in situ* etching approach may pave the way for chirality-controlled growth of SWCNTs.

3. METHODS

3.1 Preparation of the partially carbon-coated Co nanoparticles

A BCP self-assembly technique was used to prepare Co nanoparticles with a uniform diameter[53,54]. Briefly, 0.25 wt% PS-b-P4VPs (Mn: 40,000 g mol^{-1} for PS and 5,600 g mol^{-1} for P4VP, Polymer Source Inc.) were dissolved in a toluene/THF=3/1 (wt ratio) solution and electromagnetically stirred for 2 h in a 70 °C water bath. The solution was then spin-coated (Spin Master100) onto a Si/SiO_2 (300 nm thick) substrate at 5,000 r.p.m. for 2 min. The substrate had previously been treated in a sulfuric acid/hydrogen peroxide (5/1 in wt ratio) solution and ultrasonicated in ethanol and acetone for 15 min. The BCP thin films were then annealed under a saturated vapour of toluene/THF (1/4 vol ratio) solution, after which they were immersed in a 1 mmol $^{l-}$ $K_3[Co(CN)_6]$ (98%, J&K) acidic aqueous solution for 3 min, to allow the adsorption of$[Co(CN)_6]^{3-}$, followed by washing three times in de-ionized water. The BCP films were then dried at 60 °C for 20 min followed by exposure to air plasma for 10 min, to remove the polymer covered on the top of particles. The CoO nanoclusters were reduced under a hydrogen atmosphere with a flow rate of 200 s.c.c.m. at 500 °C for 5 min, subsequently heated up to 700 °C for 5 min, to obtain Co nanoparticles that were partially coated with carbon.

3.2 Synthesis of SWCNTs

For the growth of SWCNTs, a Si/SiO_2 substrate with catalyst dispersed on the surface was placed at the low-temperature zone (end) of a quartz tube reactor (25 mm in diameter) inserted into a horizontal tube furnace. When the furnace temperature reached 700 °C, the Si substrate was put into the centre of the reactor. At the same time, a 75 s.c.c.m. argon flow through an ethanol bubbler (in a 35 °C water bath) and a 200 s.c.c.m. hydrogen flow were introduced into the reactor for 10–25 min, to grow SWCNTs. The furnace was then cooled to room temperature under the protection of the Ar/H_2 flow.

3.3 Characterizations of the catalysts and SWCNTs

An AFM (Digital Instruments Multiple Mode SPM Nanoscope IIIa, operated at tapping-mode), an SEM (FEI XL30 S-FEG, operated at 1 kV), a TEM, (FEI Tecnai F20, 200 kV; JEOL JEM-2200FS 2 × CS corrected TEM, 80 kV) and a micro-Raman spectroscope (Jobin Yvon HR800, excited by 488, 532, 633 and 785 nm He-Ne laser with laser spot size of ~1 μm^2, in line mapping mode) were used to characterize the SWCNTs. For Raman characterization, more than 40 locations in each sample were measured randomly by moving the laser spot in 10 µm steps. For TEM observation, we used the transfer procedure developed by Jiao *et al.*[55, 56] with a slight modification. Briefly, a polymethylmethacrylate (PMMA) solution (AR-P 679, molecular weight=950 K, 4 wt.% in ethyl lactate, Allresist) was first spin-coated (3,000 r.p.m. for 1 min) onto the substrate on which the SWCNTs were grown. The substrate was then baked at 80 °C for 4 h in an oven. The PMMA solidified and formed a thin film with SWCNTs embedded in it. It was then put into a freshly prepared NaOH aqueous solution (5 M) with a temperature of 80 °C and kept there for 100 min. The slight etching of the SiO_2 surface by the hot NaOH solution released the PMMA film from the substrate[57]. Finally, the PMMA film embedded with SWCNTs was peeled off and attached to a Cu grid for TEM characterization after the PMMA was removed in an acetone bath[58]. For ultraviolet/visible/near-infrared absorption characterization, total 125 Si substrates (10 mm × 10 mm) with SWCNTs grown were inmmersed into a 5-ml 1 wt% sodium benzenesulfonate (SDS) D_2O solution (one by one), followed by ultrasonic treatment for 5 min, to detach the SWCNTs from the substrate and to disperse in the D_2O solution.

3.4 Calculation of the s-SWCNT content from absorption spectrum

According to the tight-binding model[34], the band gaps of s-SWCNTs and m-SWCNTs are dependent on their diameters as follows[38]:

$$E_{11}^{S}=\frac{2a_0\gamma_0}{d_t},\ E_{22}^{S}=\frac{4a_0\gamma_0}{d_t},\ E_{33}^{S}=\frac{8a_0\gamma_0}{d_t},\ E_{11}^{M}=\frac{6a_0\gamma_0}{d_t} \qquad (1)$$

a_0 is the C–C band distance and γ_0 is interaction energy between neighbouring C atoms.

In this study, the diameter of the SWCNTs is distributed in the range of 1.6–1.9 nm. Thus, their absorption spectra peaks of S_{11}, S_{22}, S_{33} and M_{11} are calculated to be at 2,408–2,860 nm, 1,204–1,430 nm, 602–715 nm and 803–953 nm, respectively. The absorption peaks of S_{11} cannot be detected due to the strong noise absorption from the solvent at corresponding wavelength. Therefore, the content of s-SWCNTs was calculated by using absorption peak areas of M_{11} andS_{22} with the following formula:

$$R_S=\frac{n_S}{n_M+n_S}=\frac{1}{1+\frac{M_{11}}{S_{22}f}} \qquad (2)$$

Where R_S is the content of s-SWCNTs in the sample, n_M and n_S are the numbers of m- and s-SWCNTs, M_{11} and S_{22} are the areas of the M_{11} or S_{22} peaks measured, and f is the absorption coefficient, here 1.24 according to the SWCNT diameter distribution. After baseline subtraction, the peak areas of S_{22} and M_{11} were measured to be 3.81 and 0.0265, respectively. Thus, the content of s-SWCNTs in our samples was calculated to be ~99% (ref. 39).

3.5 Fabrication of SWCNT TFTs

Bottom-gate TFTs based on our as-grown SWCNT network were fabricated on Si substrates with a thermally grown SiO_2 layer (300 nm) as a gate dielectric. The bottom-gate electrode (Ti/Au: 10/50 nm) was deposited by electron-beam evaporation after the SiO_2 layer on the backside of the wafer was etched by reactive ion etching. Source and drain electrodes (Ti/Au: 10/50 nm) were fabricated by standard photolithography, electron-beam evaporation and lift-off processes. Subsequently, SWCNTs outside the channel area were removed by oxygen plasma. For the fabrication of TFTs with short channel lengths, the patterning of electrodes and channel was realized by EBL. For details, the source and drain electrodes (Ti/Au: 5/50 nm) on Si/SiO_2substrates with grown SWCNT network were fabricated by EBL, electron-beam evaporation and lift-off processing. Next, the SWCNTs outside the channel area were removed by EBL and oxygen plasma etching. Electrical measurements were conducted using a semiconductor analyser (Agilent B1500A) at ambient conditions.

3.6 Preparation of exposed Co catalyst and growth of SWCNTs

For comparison, a parallel experiment was conducted by using completely exposed Co nanoparticles as the catalyst to grow SWCNTs. The catalyst was prepared by omitting the solvent annealing step to obtain a uniform BCP film without microphase separation. We treated the material in air at 750 °C for 5 min to remove the carbon coating, to obtain uncoated Co nanoparticles. The fully exposed Co catalyst was used to grow SWCNTs with the same CVD conditions described in Methods.

3.7 Synthesis of SWCNTs with a tunable mean diameter

To tune the mean diameter of SWCNTs, we prepared carbon-coated Co catalyst with a larger area of exposed Co by changing the thermal reduction process from 500 °C for 5 min to 700 °C for 10 min and to 800 °C for 10 min under H_2 with a flow rate of 200 s.c.c.m. The catalyst was used to grow SWCNTs with identical CVD growth conditions as described in Methods. The result of Raman characterization is shown in Supplementary Fig. 9. These SWCNTs have a mean diameter of 2.1 nm and the content of semiconducting nanotubes was calculated to be higher than 96%.

REFERENCES

1. Frank, S., Poncharal, P., Wang, Z. L. & de Heer, W. A. Carbon nanotube quantum resistors. *Science* **280**, 1744–1746 (1998).

2. Saito, R., Fujita, M., Dresselhaus, G. & Dresselhaus, M. S. Electronic-structure of chiral graphene tubules. *Appl. Phys. Lett.* **60**, 2204–2206 (1992).

3. Durkop, T., Getty, S. A., Cobas, E. & Fuhrer, M. S. Extraordinary mobility in semiconducting carbon nanotubes. *Nano Lett.* **4**, 35–39 (2004).

4. Tulevski, G. S. *et al.* Toward high-performance digital logic technology with carbon nanotubes. *ACS Nano* **8**, 8730–8745 (2014).

5. Cao, Q. & Rogers, J. A. Ultrathin films of single-walled carbon nanotubes for electronics and sensors: a review of fundamental and applied aspects. *Adv. Mater.* **21**, 29–53 (2009).

6. Sun, D. M. *et al.* Flexible high-performance carbon nanotube integrated circuits. *Nat. Nanotechnol.* **6**, 156–161 (2011).

7. Liu, H. P., Nishide, D., Tanaka, T. & Kataura, H. Large-scale single-chirality separation of single-wall carbon nanotubes by simple gel chromatography. *Nat. Commun.* **2**, 309 (2011).

8. Lee, H. W. *et al.* Selective dispersion of high purity semiconducting single-walled carbon nanotubes with regioregular poly(3-alkylthiophene)s. *Nat. Commun.* **2**, 541 (2011).

9. Tu, X. M., Manohar, S., Jagota, A. & Zheng, M. DNA sequence motifs for structure-specific recognition and separation of carbon nanotubes. *Nature* **460**, 250–253 (2009).

10. Zheng, M. *et al.* DNA-assisted dispersion and separation of carbon nanotubes. *Nat. Mater.***2**, 338–342 (2003).

11. Chiang, W. H. & Sankaran, R.M. Linking catalyst composition to chirality distributions of as-grown single-walled carbonnanotubes by tuning Ni_xFe_{1-x} nanoparticles. *Nat. Mater.* **8**,882–886 (2009).

12. Zhang, S. C., Tong, L. M., Hu, Y., Kang, L.X. & Zhang, J. Diameter-specific growth of semiconducting SWNT arrays using uniform Mo_2C solid catalyst. *J. Am. Chem. Soc.* **137**,8904–8907 (2015).

13. Li, Y. M. *et al.* On the origin of preferential growth of semiconducting single-walled carbon nanotubes. *J. Phys. Chem. B* **109**, 6968–6971 (2005).

14. Zhang, G. Y. *et al.* Selective etching of metallic carbon nanotubes by gas-phase reaction.*Science 2006* **314**, 974–977 (2005).

15. Hong, G. *et al.* Direct growth of semiconducting single-walled carbon nanotube array. *J. Am. Chem. Soc.* **131**, 14642–14643 (2009).

16. Ding, L. *et al.* Selective growth of well-aligned semiconducting single-walled carbon nanotubes. *Nano Lett.* **9**, 800–805 (2009).

17. Yu, B. *et al.* Bulk synthesis of large diameter semiconducting single-walled carbon nanotubes by oxygen-assisted floating catalyst chemical vapor deposition. *J. Am. Chem. Soc.* **133**, 5232–5235 (2011).

18. Li, W. S. *et al.* High-quality, highly concentrated semiconducting single-wall carbon nanotubes for use in field effect transistors and biosensors. *ACS Nano* **7**, 6831–6839(2013).

19. Qin, X. J. *et al.* Growth of semiconducting single-walled carbon nanotubes by using ceria as catalyst supports. *Nano Lett.* **14**, 512–517 (2014).

20. Harutyunyan, A. R. *et al.* Preferential growth of single-walled carbon nanotubes with metallic conductivity. *Science* **326**, 116–120 (2009).

21. Yang, F. *et al.* Chirality-specific growth of single-walled carbon nanotubes on solid alloy catalysts. *Nature* **510**, 522–524 (2014).

22. Yang, F. *et al.* Growing zigzag (16,0) carbon nanotubes with structure-defined catalysts. *J. Am. Chem. Soc.* **137**, 8688–8691 (2015).

23. Scott, L. T. *et al.* A short, rigid, structurally pure carbon nanotube by stepwise chemical synthesis. *J. Am. Chem. Soc.* **134**, 107–110 (2012).

24. Fort, E. H., Donovan, P. M. & Scott, L. T. Diels-Alder reactivity of polycyclic aromatic hydrocarbon bay regions: implications for metal-free growth of single-chirality carbon nanotubes. *J. Am. Chem. Soc.* **131**, 16006–16007 (2009).

25. Ding, F., Harutyunyan, A. R. & Yakobson, B. I. Dislocation theory of chirality-controlled nanotube growth. *Proc. Natl Acad. Sci. USA* **106**, 2506–2509 (2009).

26. Artyukhov, V. I., Penev, E. S. & Yakobson, B. I. Why nanotubes grow chiral. *Nat. Commun.* **5**, 4892 (2014).

27. Sanchez-Valencia, J. R. *et al.* Controlled synthesis of single-chirality carbon nanotubes.*Nature* **512**, 61–64 (2014).

28. Fiawoo, M. F. C. *et al.* Evidence of correlation between catalyst particles and the single-wall carbon nanotube diameter: a first step towards chirality control. *Phys. Rev. Lett.* **108**,195503 (2012).

29. Yao, Y. G. *et al.* Temperature-mediated growth of single-walled carbon nanotube intramolecular junctions. *Nat. Mater.* **6**, 283–286 (2007).

30. Mai, Y. Y. & Eisenberg, A. Self-assembly of block copolymers. *Chem. Soc. Rev.* **41**,5969–5985 (2012) .

31. Lu, J. *et al.* Selective interaction of large or charge-transfer aromatic molecules with metallic single-wall carbon nanotubes: critical role of the molecular size and orientation. *J. Am. Chem. Soc.* **128**, 5114–5118 (2006).

32. Strano, M.S. *et al.* Electronic structure control of single-walled carbon nanotube functionalization. *Science* **301**, 1519–1522 (2003).

33. Zhu, Z., Jiang, H., Susi, T., Nasibulin, A. G. & Kauppinen, E. I. The use of NH_3 to promote the production of large-diameter single-walled carbon nanotubes with a narrow (n,m) distribution. *J. Am. Chem. Soc.* **133**, 1224–1227 (2011).

34. Kataura, H. *et al.* Optical properties of single-wall carbon nanotubes. *Synth. Met.* **103**,2555–2558 (1999) .

35. Zhang, D. Q. *et al.* (n,m) assignments and quantification for single-walled carbon nanotubes on SiO_2/Si substrates by resonant Raman spectroscopy. *Nanoscale* **7**,10719–10727 (2015) .

36. Kang, L. X. *et al.* Growth of close-packed semiconducting single-walled carbon nanotube arrays using oxygen-deficient TiO_2 nanoparticles as catalysts. *Nano Lett.* **15**, 403–409(2015).

37. Jorio, A. *et al.* G-band resonant Raman study of 62 isolated single-wall carbon nanotubes.*Phys. Rev. B* **65**, 1554121 (2002).

38. Saito, R., Dresselhaus, G. & Dresselhaus, M. S. Trigonal warping effect of carbon nanotubes. *Phys. Rev. B* **61**, 2981–2990 (2000).

39. Huang, L. *et al.* A Generalized method for evaluating the metallic-to-semiconducting ratio of separated single-walled carbon nanotubes by UV-vis-NIR characterization. *J. Phys. Chem. C* **114**, 12095–12098 (2010).

40. Jiang, H., Nasibulin, A. G., Brown, D. P. & Kauppinen, E. I. Unambiguous atomic structural determination of single-walled carbon nanotubes by electron diffraction. *Carbon* **45**,662–667 (2007).

41. Tans, S. J., Verschueren, A. R. M. & Dekker, C. Room-temperature transistor based on a single carbon nanotube. *Nature* **393**, 49–52 (1998).

42. Martel, R., Schmidt, T., Shea, H. R., Hertel, T. & Avouris, P. Single- and multi-wall carbon nanotube field-effect transistors. *Appl. Phys. Lett.* **73**, 2447–2449 (1998).

43. LeMieux, M. C. *et al.* Self-sorted, aligned nanotube networks for thin-film transistors.*Science* **321**, 101–104 (2008).

44. Wang, C. *et al.* Wafer-scale fabrication of separated carbon nanotube thin-film transistors for display applications. *Nano Lett.* **9**, 4285–4291 (2009).

45. LeMieux, M. C. *et al.* Solution assembly of organized carbon nanotube networks for thin-film transistors. *ACS Nano* **3**, 4089–4097 (2009).

46. Arnold, M. S., Green, A. A., Hulvat, J. F., Stupp, S. I. & Hersam, M. C. Sorting carbon nanotubes by electronic structure using density differentiation. *Nat. Nanotechnol.* **1**, 60–65(2006).

47. Cao, Q. *et al.* Medium-scale carbon nanotube thin-film integrated circuits on flexible plastic substrates. *Nature* **454**, 495–500 (2008).

48. Cao, X. *et al.* Screen printing as a scalable and low-cost approach for rigid and flexible thin-film transistors using separated carbon nanotubes. *ACS Nano* **8**, 12769–12776 (2014).

49. Chen, H. T., Cao, Y., Zhang, J. L. & Zhou, C. W. Large-scale complementary macroelectronics using hybrid integration of carbon nanotubes and IGZO thin-film transistors. *Nat. Commun.* **5**, 4097 (2014).

50. Liu, W. W., Aziz, A., Chai, S. -P., Mohamed, A. R. & Hashim, U. Synthesis of single-walled carbon nanotubes: effects of active metals, catalyst supports, and metal loading percentage. *J. Nanomater.* **2013**, 592464 (2013).

51. Liu, B. L. *et al.* Chirality-dependent reactivity of individual single-walled carbon nanotubes.*Small* **9**, 1379–1386 (2013).

52. Li, J. H. *et al.* Importance of diameter control on selective synthesis of semiconducting single-walled carbon nanotubes. *ACS Nano* **8**, 8564–8572 (2014).

53. Mun, J. H. *et al.* Monodisperse pattern nanoalloying for synergistic intermetallic catalysis.*Nano Lett.* **13**, 5720–5726 (2013).

54. Park, S., Kim, B., Wang, J. Y. & Russell, T. P. Fabrication of highly ordered silicon oxide dots and stripes from block copolymer thin films. *Adv. Mater.* **20**, 681–685 (2008).

55. Jiao, L. Y. *et al.* Creation of nanostructures with poly(methylmethacrylate)-mediated nanotransfer printing. *J. Am. Chem. Soc.* **130**, 12612–12613 (2008).

56. Jiao, L. Y., Xian, X. J. & Liu, Z. F. Manipulation of ultralong single-walled carbon nanotubes at macroscale. *J. Phys. Chem. C* **112**, 9963–9965 (2008).

57. Reina, A. *et al.* Transferring and identification of single- and few-layer graphene on arbitrary substrates. *J. Phys. Chem. C* **112**, 17741–17744 (2008).

58. Wang, Y. *et al.* Optimizing single-walled carbon nanotube films for applications in electroluminescent devices. *Adv. Mater.* **20**, 4442–4449 (2008).

5

VIBROELECTRONIC PROPERTIES OF FUNCTIONALIZED SINGLE-WALLED CARBON NANOTUBES AND DOUBLE-WALLED BORON NITRIDE NANOTUBES

Metin Aydin[1] and Daniel L. Akins[2]

[1] Department of Chemistry, Faculty of Art and Sciences, Ondokuz Mayıs University, Samsun, Turkey

[2] Center for Analysis of Structures and Interfaces (CASI), Department of Chemistry, The City College of The City University of New York, New York, USA

1. INTRODUCTION

Carbon is the first element in group-IVof the periodic table and has a $1s^2 2s^2 2p^2$ electronic configuration, in which four valence electrons allow it to form a number of so-called hybridized atomic orbitals. Carbon atoms in elemental substances bond to each other covalently by the sharing of electron pairs, in which the covalent bonds have directional properties; this in turn provides carbon the capability to form various molecular and crystalline solid structures. The nature of the covalent bonds that are formed dictate the varied chemical and physical properties of carbon allotropes. Pure carbon-based materials not only exist as the commonly recognizeddiamond and graphite allotropes, but also more exotic entities such as fullerenes, carbon nanotubes (CNTs), and graphene; these latter allotropes having proven themselves important materials in nanotechnology.

The present chapter deals with single-walled carbon nanotubes (SWNTs), whose unique properties, as suggested above, derive from their distinctive structure. In SWNTs the carbon bonding that exists is akin to that that exists in graphite as opposed to that found in diamond. More specifically, diamond has a coordination number of four, with sp^3 hybridization, while, on the other hand, sp^2 hybridization exists in the planar layers of carbon atoms that give graphite its structure, and in the bonding that leads to the tubular structure of SWNTs. The sp^2 hybridization in graphite links carbon atoms in a two-dimensional (2D) layer of hexagons that lead to each layer in the graphite structure, in the ideal case, forming a planar structure. Each carbon atom contributes 3 electrons to 3 equivalent sigma bonds within the plane and has 1 electron left in the perpendicular p_z orbitals; such electrons are delocalized over the entire plane, resulting in a π-electron orbital system that allows the fourth valence electron to essentiallymove freely over the plane. Within the layers, the carbon-carbon bond distance is similar to the bond length in benzene (i.e., the carbon atoms are strongly bound to each other and the carbon-carbon distance is about 0.14 nm), leading to a very large inplane value for Young's modulus. However, the distance between layers (ca., 0.34 nm) is sufficiently large that the layers are bounded to each other mainly by weak, long-range Van der Waals type interaction. The weak interlayer coupling gives graphite the property of a seemingly very soft material, a property that makes graphite suitable for use in pencils and in lubricants.

As a result of its intrinsic structure, the electrical conductivity of graphite is directionally-dependent. Delocalized π-electrons parallel to the planes essentially experience metallic conduction, while electron mobility perpendicular to the layered planes would typically be much lower, but with possibly significant temperature dependency, thereby imbruing graphite as semiconductor character as well. The directionality of the conductivity translates to a band structure that has a filled valence bandand an empty conduction band separated by an energy gap. These bands, in one picture, would result from bonding and antibonding molecular π-orbitals that can be conceptualized in terms of energy lowering and energy raising combination of the perpendicular p_z atomic orbitals. The π-bonding orbitals would be fully occupied while the π-antibonding orbitals would be unoccupied, with the gap being the energy difference between the top and bottom of the respective orbitals. Because of the larger distance between its layers, graphite may form intercalation compounds with added species that act as electron donors, with graphite acting as an electron acceptor, incorporating the donated electrons into the vacant conduction band; or as electron acceptors, where graphite donates electrons from the full valence band. In diamond, it is to be noted that all valence electrons are localized around the carbon atoms, hence, such a structural characteristic has profound effectson its electrical properties, with diamond being an insulator with a band gap around 6 eV.

We now move to a more focused discussion of carbon nanotubes. Carbon nanotubes (CNTs) were discovery in 1991 [1], their unique physical, chemical, and electronic properties have led to a variety of technological application in functional nanodevices, especially as transistors and sensors [2, 3, 4], [5,6, 7], in heat conduction systems [8, 9], in specialty electronics [10, 11], molecular memories [12], optics [13, 14, 15], electrically excited single-molecule light sources [16, 17, 18, 19], to functionalized DNA [20, 21], high-performance adsorbent electrode material for energy-storage device [22], and protein functionalization [23, 24].

As it is well known, carbon nanotubes can be obtained by rolling up a defined projected area from within the hexagonal lattice of a graphene sheet in a seamless fashion such that all carbon–carbon (C–C) valences are satisfied, and the direction in which the roll up is performed transforms into the circumference of the tube. The projected area is in fact a homomorphic representation of a particular carbon nanotube [48(f-g)]. The roll-up vector is also termed the chiral vector, and is defined asna⃗ 1+ ma⃗ 2na→1+ ma→2, where a⃗ 1a→1 and a⃗ 2a→2 are the unit vectors of the hexagonal lattice, and n and m are the so-called chiral indices. An infinite number of nanotube geometries are possible, with a specific nanotube characterized by chiral indices (n,m), which, in turn, define the chiral angle θ and tube diameter (d_t); the latter is also dependent on the C–C bond length of the hexagonal lattice. For n = m, the nanotube is said to have the "armchair" conformation; for n≠0n≠0and m = 0, the conformation is called "zigzag"; while for n≠0n≠0and m≠0m≠0the conformation is termed "chiral."The diameter of the nanotube normally has values that range up to several nanometers from ~0.4 nm, while nanotubes are usually several microns in length. It is to be noted that single-walled and multi-walled carbon nanotubes generally have properties that are significantly different, while double-walled carbon nanotubes (DWNTs) can be viewed as representing the key structure that defines the transition between SWNTs and MWNTs.

Carbon nanotubes can be metallic or semiconducting depending on their structure. This is due to the symmetry and the unique electronic structure of graphene. If the chiral indices are equal, n = m, the nanotube is metallic; if n–m is a multiple of 3, then the nanotube is semiconducting, with a very small band gap; otherwise, the nanotube is a moderate semiconductor [25]. Interestingly, some nanotubes have conductivities higher than that of copper, while others behave more like silicon.

As it is well known, the optical properties of nanotubes are implicitly connected with the absorption, photoluminescence, and Raman spectroscopy of nanotubes. Such optical measurements permit a reliable characterization of the quality of nanotube, such as chirality, size, and structural defect. In the case of Raman measurements, even though a large number of phonon modes of carbon nanotubes would be expected, most of them are Raman inactive due to the selection rules that emanate from the high symmetry properties of the nanotubes. The Raman spectrum of a carbon nanotube exhibits a few characteristic modes that can be used to determine the size of nanotubes and to classify the type of the nanotubes, such as semiconducting and metallic. For example, in the low frequency region, one type of characteristic vibration is called the radial breathing mode (RBM); this movement of the carbon atoms is in the radial direction with the same phase, and corresponds to vibration of the entire tube, which is strongly diameter dependent [48(f-g)]. The RBM gives precise information about the nanotube diameter and is typically found between 100 cm^{-1} and 500 cm^{-1}. Additionally, in the high energy range from (1000 to 2000 cm^{-1}), there are two important characteristic Raman bands: the defect induced disordered band (D-band) that appears between 1300 and 1400 cm^{-1}, and tangential modes (G-band) that lie in the range from ~1560 to ~1600 cm^{-1}. The D-band is present in all graphite-like carbons and originates from structural defects. Therefore, the intensity ratio of the G/D modes is conventionally used to quantify the structural quality of carbon nanotubes. The G-band corresponds to planar vibrations of carbon atoms and is present in most graphite-like materials (at around 1580 cm^{-1}). This tangential mode (G-band) in SWCNT is split into several peaks. The splitting pattern and intensity depend on the tube structure and excitation energy; they can be used, though with much lower accuracy compared to RBM mode, to estimate the tube diameter and whether the tube is metallic or semiconducting. [48(f-g)].

Chemical functionalization by bond formation or by coating the nanotubes with organic/inorganic molecules or by encapsulating a varieties of semiconductor particles, including CdSe and CdTe, may lead to efficient energy transfer between the molecules and nanotube, as well as lead to significant enhancement in the optical properties of the composite [26, 27, 28, 29]. As an example of effect on an optical property, it has been reported that when a squarylium dye isencapsulated into a carbon nanotube, increased chemical and thermal stability of squarylium molecules occur, which, since encapsulation of a dye quenches strong dye luminescence, allows measurement and analysis of the dye's Raman spectra[30]. Also, L. Alvarez *et al.*[31] have reported that while infrared spectroscopy (IR) might provide evidence of a significant positive charge transfer for an inserted oligothiophene, Raman spectra evince different behaviors depending on the excitation energy and relationship to the oligomer's (specifically, quaterthiophene) optical absorption energy. For example, at high excitation wavelength (far from the oligomer's resonance), radial breathing modes exhibit a significant blue-shift as a result of the encapsulation effect, while at low excitation wavelength, close to resonance with the oligomer absorption, both the G-band and the low-frequency modes vanish, suggesting a significant charge transfer between the oligomer and the nanotube.

CNTs are also widely used in the clinical and research medical arenas. They find application as superior drug delivery media, for health monitoring devices; as biosensing platforms for the treatment of various diseases; in chemical sensor devices, etc. [32, 33, 34, 35]. Functionalized-SWNTs (i.e., f-SWNTs) have been known to increase solubility and permit efficient tumor targeting/drug delivery; prevents SWNTs from being cytotoxic; and possibly altering the functioning of immune cells. Moreover, carbon nanotubes have enhanced solubility when functionalized with lipids that make their movement through the human body easier and reduces the risk of blockage of vital body organ pathways. Also, CNTs exhibit strong optical absorbance in certain spectral windows, such as the NIR (near-infrared); when functionalized within tumor cell with specific binding entities, the nanotubes have allowed the selective destruction of disease (e.g., cancer) cells with NIR in drug delivery applications.

More recently, boron nitride nanotubes (BNNTs) can be counted among the modified CNT that have been synthesized [36, 37, 38]. The electronic properties of boron nitride nanotubes differ from carbon nanotubes: while carbon nanotubes can be either metallic or semiconducting, depending on their chirality and radius [39], all boron nitride nanotubes (BNNTs) are found to be semiconducting materials with a large band gap[40]. And since the band gap is large, the gap energy is only weakly dependent on the diameter, chirality, and the number of walls of a multi-walled tube structure. Moreover, because of their semiconducting character, BNNTs like CNTs themselves are also very interesting materials for application in nanoscale devices, and have been considered alternatives to CNTs [41, 42]. Like CNTs the modification of the electronic properties of BNNTs by doping and functionalization is an important avenue for making nanodevices. The doped BNNTs nanotubes may exhibit a dramatic change relative to the pristine nanotube. Furthermore, because of the strong interactions between electrons and holes in BNNTs [43, 44], the excitonic effects in BNNTs have proven more important than in CNTs. Bright and dark excitons in BNNTs qualitatively alter the optical response [45].

For a better understanding of the physical and optical properties of nanotubes, quantum mechanical calculations have been extremely helpful. In this chapter, we provide theoretical results on double-walled boron nitride nanotubes (DWBNNTs) and functionalized nanotubes using DFT; this report extends the quantum chemical computational approach that we have used earlier [48(f-g)]. The results of calculations not only indicate the shift in the spectral peak positions of the RBM and G-modes in Raman spectra of DWBNNTs relative to their corresponding isolated SWBNNTs, but also indicatesa charge transfer from the outer-shell to the inner-shell when DWBNNTs are excited, as discussed in Section 3. Furthermore, the plots of the frequencies of vibrational radial breathing modes (RBM) versus $1/d_t$ for (2n,0)&(n,0)-DWBNNTs exhibit a strong diameter dependence.

For functionalized single-walled carbon nanotubes, we find that there should be a charge transfer process directed from the nanotube to an attached molecule, which is active in optical excitations. More generally, upon irradiation a system can undergo internal conversion (IC) and intersystem crossing (ISC) processes, in addition to photochemical and other photophysical processes. Transient intermediates are likely to form in the IC and ISC radiationless processes, herein referred to as "dark processes," which are not detected using conventional light absorption or emission spectroscopic methods. As seen from the combined Perrin-Jablonski diagram in Figure 1, for a typical molecule the emission of a photon from an electronically excited state to the ground state results in fluorescence in the region of 300 to 1500 nm. Photophysical processes for an isolated molecule occur as a result of transitions between the different internal energy states of that comprise the electronic states. A molecular system in the gas-phase or in the solution phase at room temperature is mostly expected to be in its ground state (S_0). The excitation of a molecular system from its ground state to an excited vibroelectronic state by absorption of a photon (occurring within ca. 10^{-15} second) is much faster than a emission of the photon from its excited electronic state (S_k, k>1) to its ground state (occurring in ca. 10^{-8} second). All of the excited molecular systems may not directly return back to their ground state by emission of a photon,

$S_{k>0} \rightarrow\rightarrow S_0$ transition, but some of them may return back to their ground states (S_0) by internal conversion (CI), for instance, when the molecule is excited into a higher vibroelectronic state (S_k>1), it may undergo relaxation to the S_1 state (in 10^{-12} s) via vibrational coupling between these states before undergoing additional vibrational relaxation and returning to the lowest singlet electronic energy level (S_1), referred to as internal conversion. Subsequently, transition from S_1 to S_0 by emission of a photon (fluorescence) occurs. An alternate pathway for a molecule in the lowest energy S_1 state involves intersystem crossing (at rates that can compete with fluorescence) by the molecule into a triplet state T_1. From T_1, the molecule can undergo radiative de-excitation via a much slower process, which is known as phosphorescence ($T_1 \rightarrow\rightarrow S_0$ transition), such as illustrated by the Perrin-Jablonski diagram given in Figure 1.

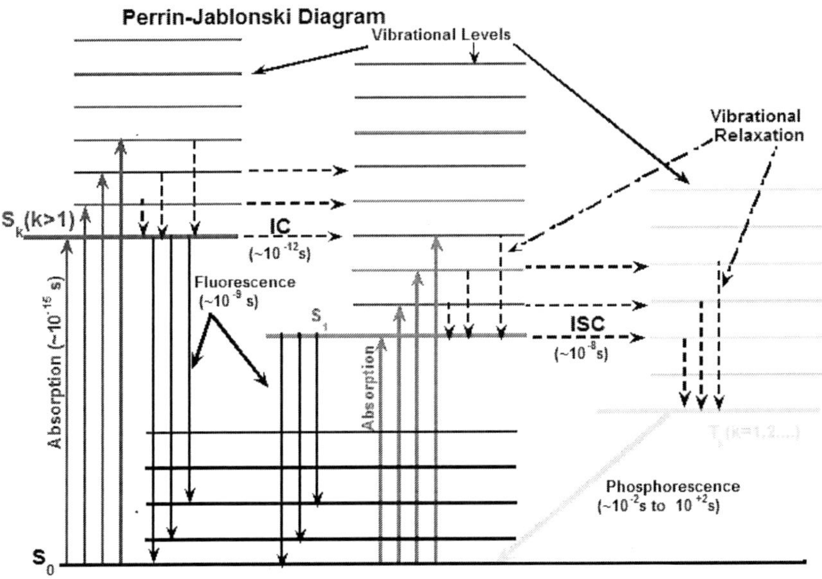

Figure 1. A general Perrin-Jablonski diagram for a fluorescent molecule, where S and T stand for singlet and triplet electronic states, respectively. IC and ISC represent "internal conversion" and "intersystem crossing", respectively.

It is to be noted that fluorescence resonance energy transfer (FRET) can be used to investigate intra- and/or intersystem energy transfer dynamics that might occur as one transitions fromsingle-walled nanotubes (SWNTs) to multi-walled nanotubes (MWNTs),or to the functionalized nanotubes (f-NTs). Such dark intermediates are expected to play crucial roles in IC and ISC processes and thus are fundamental to understanding mechanistic photochemistry of the functionalized-nanotubes and multi-walled nanotubes. We have used time-dependent DFT (i.e., TD-DFT) methods to determine the dark transient structures involved in radiationless processes for functionalized-SWCNTs and DWBNNTs. Also, we have calculated all possible singlet-triplet vertical electronic transitions and discussed these in terms of IC and ISC processes.

It is to be noted that CNTs have been shown to exhibit strong optical absorbances in certain spectral windows, such as the NIR (near-infrared). Moreover, when functionalized with tumor cell specific binding entities CNTs have facilitated the selective destruction of disease cells (e.g., cancer cell) in the NIR and play a significant role in drug delivery applications[46]. In the present chapter, we acknowledge the importance of calculating the IR spectra of both functionalized-SWCNTs and DWBNTS.

2. RESULTS AND DISCUSSION

Computational methods: The ground state geometries of single-walled carbon nanotubes (SWCNTs), double-walled carbon nanotubes (DWCNTs), single-walled boron nitride nanotubes (SWBNNTs), and functionalized-SWCNTs were optimized without symmetry restriction on the initial structures. Both structure optimization and vibrational analysis calculations were implemented using DFT with functionals, specifically, B3LYP, in which the exchange functional is of Becke's three parameter type, including gradient correction, and the correlation correction involves the gradient-corrected functional of Lee, Yang and Parr. The basis set of split valence type 6-31G, as contained in the Gaussian 03 software package[47], was used. The results of the calculations did not produce any imaginary frequencies. The vibrational mode descriptions were made on the basis of calculated nuclear displacements using visual inspection of the animated normal modes (using GaussView03) [47], to assess which bond and angle motions dominate the mode dynamics for the nanotube. The DFT method was chosen because it is computationally less demanding than other approaches as regards inclusion of electron correlation. Moreover, in addition to its excellent accuracy and favorable computation expense ratio, the B3LYP calculation of Raman frequencies has shown its efficacy in numerous earlier studies performed in this laboratory and by other researchers, often proving itself the most reliable and preferable method for many molecular species of intermediate size, including anions and cations [48]. In our calculations, hydrogen atoms have been placed at the end points of the unit cells. Furthermore, the time-dependent density functional theory at TD-B3LYP level were applied to calculate the vertical electronic transitions for the SWCNTs, SWBNNTs and functionalized (7,0)- and (10,0)-SWCNTs. For geometry optimization and calculations of electronic transitions, the 6-31G* basis set was used for sulfur atom (S) and the 6-31G basis set was used for the other atoms involved in the covalently functionalized nanotubes. It is worth nothing that the results of the calculated structural and spectroscopic properties of the double-walled boron nitride nanotubes (DWBNNTs) and the functionalized zigzag single-walled carbon nanotubes (f-(n,0)-SWCNTs) used in this chapter have been submitted to elsewhere for publication.

3. RESULTS AND DISCUSSION

3.1. Structural Results

Calculated diameters of the (0,n)&(0,2n)-DWCNTs (zigzag double-walled carbon nanotube) and (0,n)&(0,2n)-DWBNNTs (zigzag double-walled boron nitride nanotubes), for n = 6 to 10, were found to decrease for the inner-nanotube and increase for the outer-nanotube, referenced to the corresponding diameter of the zigzag single-wall nanotube ((0,n)-SWNT) which changes with n. A fit to the calculated individual tube diameters for each inner- and outer-shell of the DWCNTs and DWBNNTs using a functional form that depends inversely on single-walled nanotube's diameter: fit parameters are shown in Eq. 1a-2b

$$\begin{aligned} &D_t(\text{outer-shell-DWCNT}),\ \text{in nm}) = -0.040 + \frac{0.147}{d_t} + \frac{0.138}{d_t^2} \qquad a \\ &D_t(\text{inner-shell-DWCNT}),\ \text{in nm}) = -0.039 + \frac{0.037}{d_t} + \frac{0.005}{d_t^2} \qquad b \end{aligned} \tag{1}$$

$$D_t(\text{outer}-\text{shell}-\text{DWBNNT}),\ \text{in nm}) = -0.009 + \frac{0.114}{d_t} + \frac{0.143}{d_t^2} \quad a$$

$$D_t(\text{inner}-\text{shell}-\text{DWBNNT}),\ \text{in nm}) = -0.069 + \frac{0.081}{d_t} + \frac{0.021}{d_t^2} \quad b \qquad (2)$$

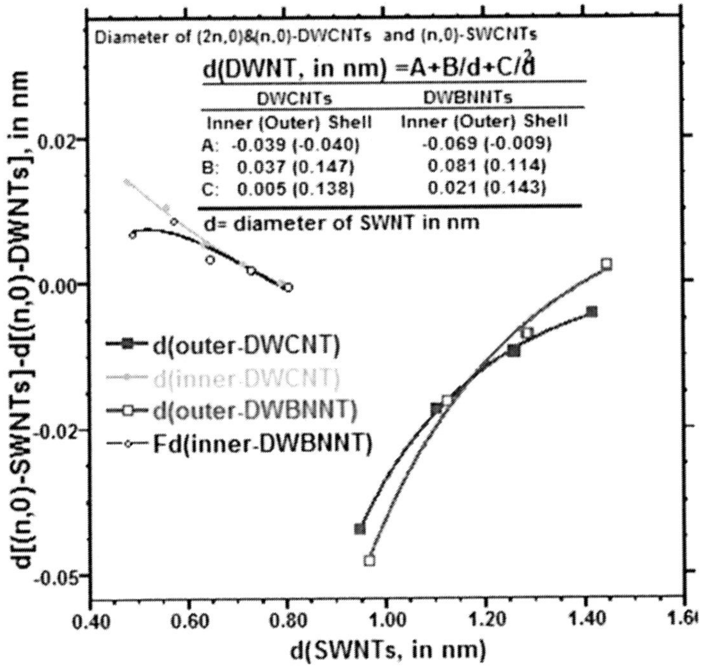

Figure 2. Calculated diameters of the double-walled carbon nanotubes, (2n,0)&(n,0)-DWCNTs, and double-walled boron nitride nanotubes, (2n,0)&(n,0)-DWBNNTs, for n = 6 to 10

A comparison the diameters of the inner- and outer-shells of the DWNTs with their corresponding SWNTs diameters show that the inner-shells diameters decrease and the outer-shells diameters increased. These predictions explicitly indicate the existence of intertube interactions in DWCNT systems. As seen in Figure 3, the diameter dependence of the curvature energies of the DWCNTs and DWBNNTs referenced to the global energies per hexagon of the (0,10)&(0,20)-DWCNTs and -DWBNNTs is well fitted by a Lennard-Jones potential expression (ELJ=−(Ar6− Br12)ELJ=-Ar6- Br12 where parameters of the are A and B are van der Waals interaction parameters in Lennard-Jones potential) as given in equation 3a-b,

$$\Delta E(\text{DWBNNTs, in eV}) = -\left(\frac{0.503}{D_t(\text{nm})}\right)^6 \left\{1-\left(\frac{0.266(\text{nm})}{D_t(\text{nm})}\right)^6\right\} \quad a$$

$$\Delta E(\text{DWCNTs, in eV}) = -\left(\frac{0.477}{D_t(\text{nm})}\right)^6 \left\{1-\left(\frac{0.356(\text{nm})}{D_t(\text{nm})}\right)^6\right\} \quad b$$

(3)

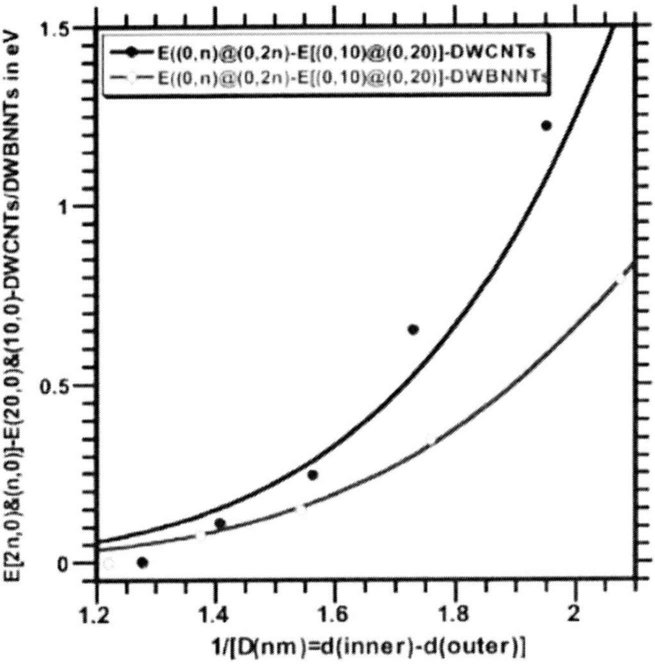

Figure 3. The diameter dependence of the curvature energies of the DWCNTs and DWBNNTs referenced to the global energies per hexagon of the (0,10)&(0,20)-DWCNTs/DWBNNTs is well fitted by a Lennard–Jones potential expression as given in Eqs. (3a) and (3b).

where $\Delta E\left(\frac{\text{DWCNTs}}{\text{DWBNNTs}}\right) = E[(2n,0)\&(n,0)] - E[(20,0)\&(10,0)]$ and $D_t = d_o(\text{outer shell}) - d_i(\text{inner shell})$ $D_t = d_t(\text{outer shell}) - d_t(\text{inner shell})$. The results of the calculations suggest that the DWNTs with large diameters can be much more easily formed than those with small diameters. When comparing the formation energy of the DWCNTs with the DWBNNTs, as shown in Figure 3, it can be seen that the formation of the DWBNNTs is favorable to that of DWCNTs due to the relatively strong interactions between the inner- and outer-shells in the case of the DWBNNTs. This finding also is supported by the calculated electron density, as discussed below, as well as the relative change in the tube diameters when going from the SWNT to the DWNT, as seen in equations 1-2. Furthermore, our ongoing calculations on the energetically stability of the DWBNNTs as function of the interwall distance (between inner- and outer-shells) indicates that the interwall distance around 0.34 nm is more stable, which are excellent agreement with the experimental observations by J. Cumings [59], which will be published elsewhere. However, at different experimental conditions, the DWBNNTs with small interwall distance such as (0,6)&(0,12)-DWBNNT might be formed at different experimentally conditions. The DWBNNTs with small interwall distance might be more interesting than other, in their optical applications.

Figures 4 A-B illustrate the calculated electron density of (12,0)&(6,0)-DWCNT and (0,n)&(0,2n)-DWBNNT, n= 6 and 8. For (12,0)&(6,0)-DWCNT, the geometry optimization, without any symmetry restriction, predicted ground state geometry has C_{2v} point group and the electronic state of ground state has singlet-A_1 symmetry. The plotted electron density showed that while first four highest occupied molecular orbitals (from HOMO to HOMO-4, of B_2, B_1 and $2E_2$ symmetries, respectively) involve both the inner- and outer-shell, the HOMO-5 with the $2E_1$ symmetry belongs to outer-shell only. The lowest unoccupied molecular orbital, LUMO (E_1) lies about 4.699 eV above the HOMO (B_2), and belongs to the inner-shell, while the next higher one (E_1) involves not only the inner- and outer-shell (lies 5.521 eV above the HOMO (A_1)), but also there is a significant sigma-bonding interaction between the inner and outer tubes in the excited state. For the (0,6)&(0,12)-DWCNT, the calculated electron density of (0,6)&(0,12)-DWCNT shows that the first four highest occupied molecular orbitals (from HOMO to HOMO-3, with the A_{1u}, A_{2g} and $2E_{1g}$ symmetries, respectively) belong to the outer-shell and the next higher occupied molecular orbitals, from HOMO-4 to HOMO-24, include both inner- and outer-shells of (0,6)&(0,12)-DWCNT. The lowest unoccupied molecular orbital, LUMO (E_{1u}) lies about 0.780 eV above the HOMO(A_{1u}) and belongs to the outer-shell, while the next one (with B_{2u}symmetry) belongs to the inner-shell, and lies 0.849 eV above the HOMO(A_{1u}). The calculated electron densities also indicate that an intratube (inner and outer tube) interaction may take place in the excited state, since the LUMO+7(A_{2u}), LUMO+8(E_{1u}), LUMO+10(E_{1g}) and LUMO+15(E_{1g}) lie about 2.494, 2.557, 2.563, 3.637 eV above the HOMO(A_{1u}), respectively.

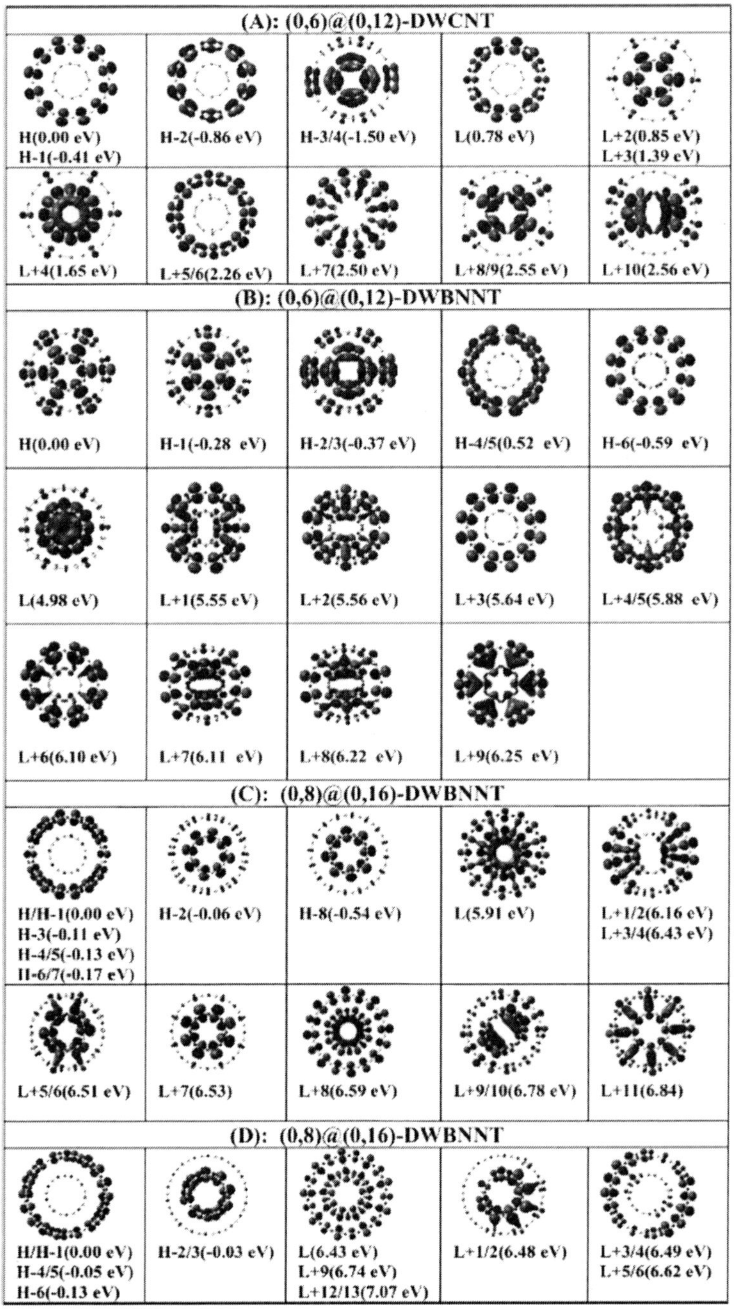

Figure 4. Calculated electron densities in the HOMO and LUMO states: A for the (0,6)&(0,12)-DWCNT, B for (0,6)&(0,12)-DWBNNT, C for (0,8)&(0,16)-DWBNNT, and D for (0,9)&(0,18)-DWBNNT.

The intratube σ-bonding interaction in the excited state of the (0,6)&(0,12)-DWBNNTs and DWCNT might lead to a probable intertube charge transfer, which can be observed by a significant change in the tangential modes (TMs) of resonance Raman spectra when the tube excited to its intratube charge transfer state. The TMs may not only provide information about the metallic or semiconducting character of nanotubes, but also about the inner-outer tube (intratube) charge transfer. Indeed, very recently, resonant Raman measurements [49], photoemission measurements, and theoretical calculations have provided evidence of charge transfer between the inner- and outer-shells of DWCNTs.

Given such a scenario, small sized-DWCNTs and DWBNNTs might be used as energy conversion systems due to charge transfer between intershells, which might be indicated by changes in of the Raman band intensities upon excitation in resonance with charge transfer between inner- and outer-shells.

3.2. Raman Spectra Of Single-Walled And Double-Walled Boron Nitride Nanotube

We calculated Raman spectra for the zigzag single-walled boron nitride nanotube ((0,n)-SWBNNTs, n=6 to 19) and double-walled boron nitride nanotube, (n,0)&(2n,0)-DWBNNTs with n=6 to 9. While the Figs. 5A and 6 provide the calculated Raman spectra for the SWBNNTs and DWBNNTs, respectively, the Figure 7 provides the Raman spectra of the (0,8)&(0,16)-DWBNNT and isolated (0,8)- and (0,16)-SWBNNTs for the comparison. Furthermore, we provided the vibrational mode assignments and frequencies for the DWBNNT and isolated SWBNNTs in Tables 1. All assignments to motions of atoms or groups of atoms in Tables 1 have been accomplished through use of vibration visualization software (specifically, GaussView03). The results of the calculations are summarized below.

Zigzag-SWBNNTs: In the low frequency region (<500 cm^{-1}), the calculated Raman spectra the (0,n)-SWBNNTs (n=6 to 19) exhibited two Raman bands. One of them is known as the radial-breathing mode (RBM) and other is elliptical deformation mode (EDM). The RBM is an important mode for the characterization and identification of particular nanotubes, especially of their chiralities. The importance of the radial-breathing mode for the characterization of nanotubes derives from the inverse dependence of its frequency on the diameter of the nanotube. As seen in Figs. 5A-B, the radial breathing mode (RBM with A_{1g} symmetry, $\omega_{RBM}(A_{1g})$) and other Raman band (elliptical deformation mode (EDM) with E_{2g} symmetry, $\omega_{EDM}(E_{2g})$ have frequencies that inversely depend on a nanotube's diameter. A linear fit to the calculate RBM frequency dependence on nanotube diameter is provided; a linear equation, $\omega_{RBM}(A_{1g}) = 48.51 + \frac{183.54\ cm^{-1}.nm}{d_t(nm)}$, which is in excellent agreement with the results of the DFT within ± 1 cm^{-1}. However, the offset constant in the linear fitting equation (48.51 cm^{-1}) produce significant error for the (0,n)-SWBNNTs with large diameter because the RBM decreases with increasing tube diameter and RBM in the limit of infinite diameter yields to a simple translation of the BN sheet. The RBM frequency should therefore go to zero in this limit. Therefore, a curve fit may be obtained using a cubic equation such as $\omega_{RBM}(cm^{-1}) = \frac{307.36\ cm^{-1}.nm}{d_t(nm)} - \frac{97.87\ cm^{-1}.nm^2}{[d_t(nm)]^2} + \frac{24.12\ cm^{-1}.nm^3}{[d_t(nm)]^3}$ which reproduces the RBMs within a ± 3 cm^{-1} error range when, comparing with the calculated Raman spectra of the SWBNNTs from (0,6) to (0,19) using the DFT technique and the RBM goes to zero in the limit of infinite diameter. An analytical expression for the other accompanying calculated low frequency bands (EDM of E_{2g} symmetry), which has lower frequency than the RBM, the best fit parameters carried out to third order in inverse diameter parameter is given by the equation: ωEDM(E2g)(=113.64+29.03 cm−1.nmdt(nm)−14.62 cm−1.nm2[dt (nm)]2+6.33 cm−1.nm3[dt(nm)]3)ωEDME2g (=113.64+ 29.03 cm-1.nmdtnm-14.62 cm-1.nm2dtnm2+6.33 cm-1.nm3dtnm3), which reproduces exact calculated values of the EDMs. It is noting worth that, without offset constant, fitting equation (linear or high order) reproduces the calculated values of the EDMs within a large error range. The band is labeled as EDM for elliptical deformation, which derives from the predominate motions that define vibrational mode motions, as ascertained with the vibration visualization software mentioned earlier.

The results of calculated Raman spectra of the (0,n)-SWBNNTs showed that: 1) the RBM of the frequency dramatically increases with decreasing the SWBNNTs diameter, which is not so surprising since the N–B–N bond strain and the sp^3 hybridization rapidly increases with decreasing SWBNNTs diameter; 2) as seen in Figure 5, for large sized SWBNNTs, the ωRBM(A1g)ωRBMA1g and ωEDM(E2g) ωEDME2g mode frequencies converge. For instance, the calculated frequency separation between the RBM and EDM is found to be 3, 7, 21 and 43 cm^{-1}, when n has the values 26, 25, 22 and 19, respectively. Thus, one can anticipate the (0, 28)-SWBNNT would have unresolvable RBM and EDM bands for the experimental spectra. We can anticipate that the acquisition of Raman spectra for experimental samples consisting of large diameter SWBNNT with the purpose of characterizing the sample in terms of electronic properties and purity may be complicated by the existence of this EDM band, which, in general, can lead to apparent broadening of bands as well as the presence of additional bands that may lead to the erroneous conclusion that more than one type of SWBNNT is present in the sample. Of course, this issue is not expected to be of great significance since the synthesis routes that are presently in vogue do not lead to nanotubes with diameter as large as that corresponding to the (0,26) index. It is to be noted that the E_{2g} band has lower frequencies than the RBM, (see Figure 5A). This latter band is labeled as EDM for elliptical deformation, which derives from the predominate motions that define vibrational mode motions, as ascertained with the vibration visualization software mentioned earlier.

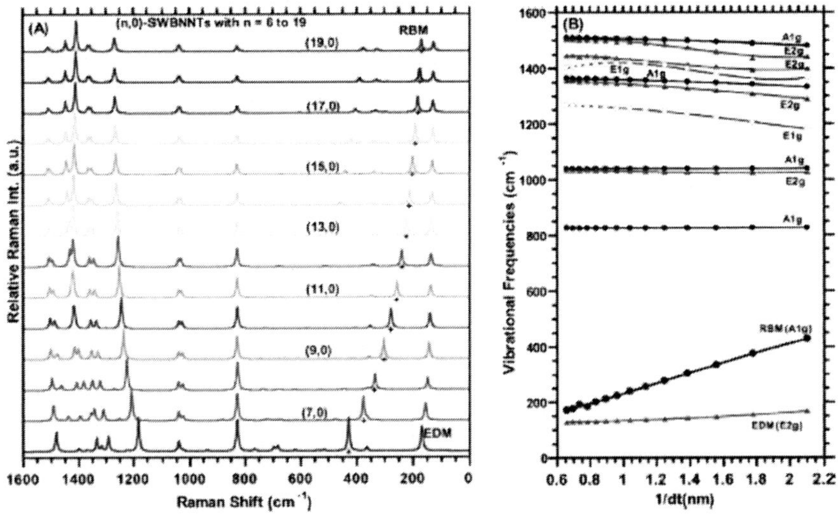

Figure 5. (A) calculated Raman spectra of the (0,n)-SWBNNTs, n = 6–19; (B) the plots of the frequencies of vibrational modes of symmetries A1g, E1g and E2g versus $1/d_t$.

Table 1. DFT-calculated Raman vibrational frequencies (in cm^{-1}) and assignments for (0,n)-SWBNNT and (0,n)&(0,2n)-DWBNNTs at the B3LYP/6-31G level.

	(0,6)	(0,7)	(0,8)	(0,9)	(0,10)	(0,11)	(0,12)	(0,13)	(0,14)	(0,15)	(0,16)	(0,17)	(0,18)	(0,19)	(0,6)& (0,12)	(0,7)& (0,14)	(0,8)& (0,16)	(0,9)& (0,18)
E_{2g}	167	155	147	143	139	137	135	134	133	131	130	130	129	128	246 156	206 147	170 139	152 130
	Elliptical deformation (EDM) of both inner and outer tubes in the same phase																	
A_{1g}	428	376	335	303	277	256	239	224	212	201	192	184	177	171	497 256	416 226	354 200	310 179
	Radial breathing of the outer tube only (RBM)																	
A_{1g}	826	827	826	827	827	827	827	827	827	827	827	827	827	828	820 831	823 832	823 833	823 832
	Out-of-surface bending deformation of the NBN/BNB bonds on the tube																	
E_{2g}	1026	1024	1025	1026	1027	1029	1030	1031	1032	1033	1033	1034	1034	1035	1030	1027	1030	1034
	BN stretching (in opposite phase) along tube axis																	
A_{1g}	1040	1039	1040	1040	1039	1040	1040	1040	1040	1040	1039	1039	1039	1039	1040	1034	1036 1036	1039 1044
	BN stretching along tube axis only																	
E_{2g}	1181	1206	1223	1236	1244	1251	1255	1258	1261	1263	1265	1266	1267	1268	1253 1234	1206 1242	1243 1246	1238 1263
	Asymmetric stretching vibrations and bending deformations of the BNB/NBN bonds																	
E_{1g}	1289	1308	1320	1329	1335	1340	1344	1347	1349	1351	1353	1354	1356	1356		1351	1341	1352
	Bending deformation of the NBN/BNB bonds, including relatively weak BN bond stretching																	
A_{1g}	1333	1343	1349	1353	1355	1358	1360	1361	1362	1363	1363	1364	1364	1365	1371	1368	1358	1363
	BN stretching and bending deformation of the NBN/BNB bonds along tube axis																	
E_{1g}	1370	1355	1382	1401	1413	1419	1419	1418	1416	1414	1412	1410	1408	1406	1366	1398	1373	1376
	Asymmetric stretching vibrations and bending deformations of the BNB/NBN bonds.																	
E_{2g}	1399	1393	1407	1414	1417	1424	1432	1437	1441	1444	1443	1446	1447	1446			1420	1429
	BN stretching and bending deformations of NBN/BNB bonds																	
E1g E2g	1442	1436	1461	1476	1486	1492	1495	1495	1499	1499	1501	1502	1503	1504	1413	1432	1434	1434
	BN stretching, including bending deformations of NBN/BNB bonds																	
A_{1g}	1481	1490	1496	1500	1502	1504	1505	1507	1507	1508	1509	1509	1509	1510	1430 1511	1463 1517	1473 1517	1485 1519
	BN stretching along tube axis, including bending deformations of NBN/BNB bonds																	

As regards other general conclusions that can be drawn from our calculations for the SWBNNTs, we have found that calculated Raman bands in the mid-frequency region exit nearly size-independent peak positions. As shown in Table 1 or Figs. 5A-B, in the high frequency region there are a few Raman bands of symmetries $E_{1g}/E_{2g}/A_{1g}$ that lie close to one another in frequency. For instance, the calculated Raman modes with symmetries of the A_{1g} (~1355 ± 10 cm^{-1}) and E_{2g} (~1330 ±25 cm^{-1}) approach one another in frequency with increasing diameter of the SWBNNT and then reach a constant values of 1365 and 1356 cm^{-1}, respectively, as seen in Table 1. A fitting equation indicated that these two Raman bands (with symmetries A_{1g} at ~ 1355 ±10 cm^{-1} and E_{2g} (~1330 ±25 cm^{-1}) first increase in frequency then approach a constant value of ~1366 and ~1360 cm^{-1}, respectively, with increasing diameter of the (0,n)-SWBNNT, n=25. Furthermore, the resonance Raman experiments [60,61] have been shown that there is only one strong band at 1355 ± 10 cm^{-1} in high energy region for the boron nitride nanotubes. Thus, the calculated these Raman bands at A_{1g} (~1355 ± 10 cm^{-1}) and E_{2g} (~1330 ±25 cm^{-1}) are not only in good agreement with experiments, but also the calculations suggest that only the Raman band(s) (of the symmetry of A_{1g} and/or E_{2g}) are theatrically enhanced by resonance excitation of the boron nitride nanotube.

Furthermore, the predicted shifts in the peak positions may result from the nanotube curvature effect as mentioned in Refs. 48(f-h), the curvature energy of the nanotube brings about dissimilar force constants along the nanotube axis and the circumference direction. Therefore, the nanotube geometry causes a force constant reduction along the tube axis compared to that in the circumferential direction. Consequently, the curvature effect might play crucial role in the shift of the peak positions of the G-band as well as the RBM band, as mentioned earlier. In addition, the calculated Raman band positions for bands at ~ 1240 ± 30 cm^{-1} are found to be slightly size dependent, exhibiting a slightly blue shift with increasing diameter of the SWBNNTs. This disorder induced mode is also important for the characterization and the defect on the nanotube as observed a broad feature around in the spectrum of the Al-modified MWBNNTs [63]. For example, in the resonance Raman enhanced spectrum, the relative intensity of the disorder mode increases relative to the intensity of the breathing and tangential modes since there is a defect on the nanotube surface as a result of chemical functionalization or caused by structural deformation. For the carbon nanotubes (CNTs), the experimental studies have showed that the increase in the intensity ratio (I_D/I_G) indicates an increase in the number of defects on the sidewall of the nanotube. This is expected result of the introduction of covalently bound moieties to the nanotube framework, in which significant amount of the sp^2 carbons is converted to sp^{3}hybridization.

DWBNNTs: While Figure 6 provides the calculated nonresonance Raman spectra for the (0,n)&(0,2n)-DWBNNTs, with n ranging from 6 to 9; Figure 8 provides diagrams of the atomic motions associated with the vibrational frequencies for the (8,0)&(16,0)-DWBNNT used as a representative case. The calculations show that the frequencies of the radial breathing modes (RBMs) and tangential modes (TMs, known as G-mode) of (n,0)&(2n,0)-DWBNNT (with n=6 to 9) significantly differ from those calculated for the (0,n)-SWBNNTs (see Figure 7 and Table 1). The results of the calculations are summarized below. In the low frequency region, the calculated Raman spectra of these DWBNNTs exhibited two RBM modes resulting from the radial motion of the inner- and outer-shells, as shown inFigure 6, and both of these RBM modes are strongly diameter dependent. A large gap between RBMs in the Raman spectra of the DWBNNTs decreases with increasing diameter of the inner- and outer-shells (as seen in Figure 6). Comparing these calculated RBMs in the spectrum of the (0,8)&(0,16)-DWBNNT with their corresponding bands in the isolated (0,8)- and (0,16)-SWBNNTs spectra, as seen in Figure 7, we note that the RBMs at 335 cm^{-1} in the Raman spectrum of the (8,0)-SWBNNT and at 192 cm^{-1} in the (16,0)-SWBNNT spectrum are, respectively, upward shifted to 354 and 200 cm^{-1} in the spectrum of (0,8)&(0,16)-DWBNNT. Additionally, the RBMs for the (0,6)-SWBNNT(428 cm^{-1}) and for the (0,12)-SWBNNT (239 cm^{-1}) spectrum are, respectively, blue shifted to 497 and 256 cm^{-1} in the Raman spectrum of (0,6)&(0,12)-DWBNNT (see Table 1). The relative distances between RBMs in the spectra of (0,n)&(0,2n)-DWCNTs are greater than the separation between corresponding RBMs in Raman spectra of (0,n)- and (0,2n)-SWCNTs. For instance, the distance between the RBMs for (0,8)&(0,16)-DWBNNT is 154 cm^{-1}, this distance between the RBMs in the Raman spectra of the corresponding isolated (0,8)- and (0,16)-SWBNNTs is 143 cm^{-1}. A tentative fitting equation may be obtained as given in Equation 4a-b:

$$\omega_{inner}(\text{RBM, in cm}^{-1}) = \frac{181.27}{d_t\,(nm)} + \frac{37.00}{[d_t\,(\text{nm})]^2} - \frac{4.82}{[d_t\,(\text{nm})]^3} \qquad a$$

$$\omega_{outer}\left(\text{RBM, in cm}^{-1}\right) = \frac{237.34}{d_t\,(nm)} + \frac{65.65}{[d_t\,(\text{nm})]^2} - \frac{51.85}{[d_t\,(\text{nm})]^3} \qquad b \qquad (4)$$

where d_t stand for the shell diameter. The tentative fitting equations reproduced calculated RBMs within 0.5 cm^{-1} error range for both inner- and outer-tubes. Another Raman bands below RBM modes in the spectra of the SWBNNTs are blue-shifted relative to the corresponding peaks in the spectra of their corresponding DWBNNTs. For instance, these Raman features at 147 cm^{-1} in the spectra of (0,8)-SWBNNT and at

130 cm^{-1} in the spectrum of the (0,16)-SWBNNT are respectively blue-shifted to 170 and 139 cm^{-1} in the spectrum of the (0,8)&(0,16)-DWBNNT. Furthermore, in the mid-frequency region, the relatively weak intense peaks are centered 1036 (A_{1g}), 1030 (E_{2g}) and 823 (A_{1g}) cm^{-1} are predicted almost at the same positions in the spectra of both (0,8)- and (0,16)-SWBNNTs.

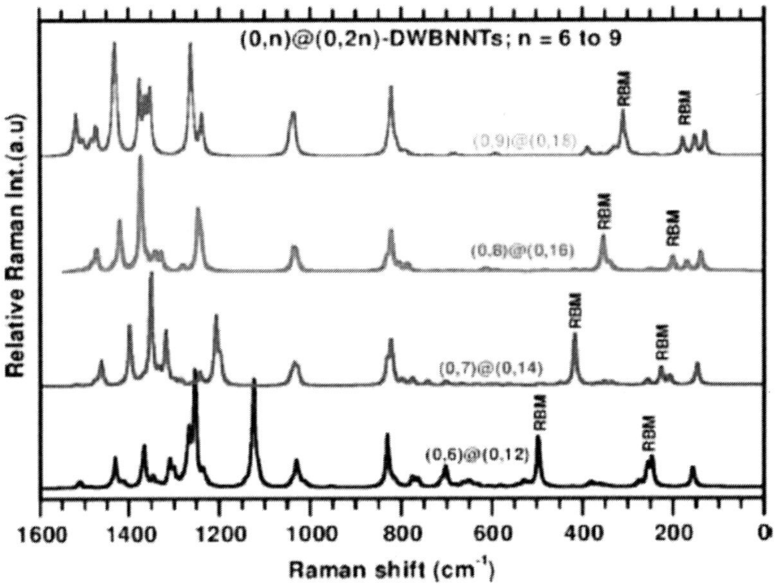

Figure 6. Calculated Raman spectra of the (0,n)&(0,2n)-DWBNNT, n = 0–9.

In the high frequency region, comparing the Raman features in the spectra of the (0,8)&(0,16)-DWBNNTs with their band position in the corresponding (0,n)-SWBNNTs spectra, it can be seen that they are slightly shifted relative to SWBNNTs, as seen in Figure 7 for the (0,8)&(0,16)-DWBNNT. For instance, the Raman bands at 1434 (A_{1g}, relatively weak), 1420 (E_{1g}, medium intense), 1373 (E_{1g}, the most stronger one), 1358 (A_{1g}, relatively weak), and 1246 (E_{1g}, relatively strong) cm^{-1} in the spectrum of the (0,8)&(0,16)-DWBNNT correspond to the Raman features at 1461 (medium), 1407 (medium), 1382 (medium), 1349 (medium), and 1223 (the most stronger) cm^{-1} in the spectrum of the (0,8)-SWBNNT, and these are predicted at 1501 (relatively weak), 1443 (medium), 1412 (the most stronger), 1363 (relatively weak), and 1265 (relatively strong) cm^{-1} in the spectrum of the (0,16)-SWBNNT (seeFigure 6), respectively.

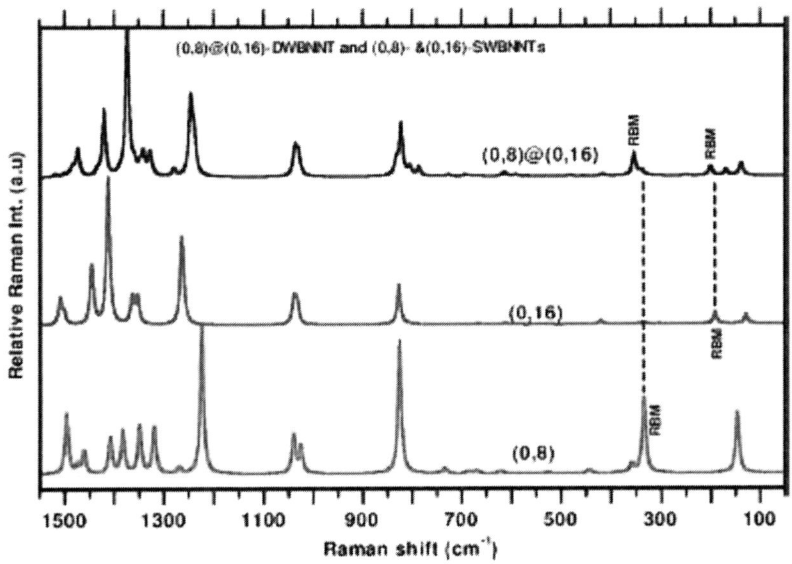

Figure 7. Calculated Raman spectra of the (0,8)&(0,16)-DWBNNT and (0,8)- and (0,16)-SWBNNTs.

Moreover, Y. Bando et. al. [62] have studied Raman spectra of the multi-walled boron (natural ^{11}B and isotope ^{10}B) nitride nanotubes (MWBNNT and MW^{10}BNNT). Their Raman spectra of the MWBNNT and MW^{10}BNNT showed only one strong Raman peak at 1366 and 1390 cm^{-1}, respectively, in the range of 1200 to 1500 cm^{-1}, which is assigned to a BN stretching deformation vibration mode. This measured Raman peak is in good agreement with our calculated Raman peak (E_{1g}) at 1373 cm^{-1} in the calculated nonresonance Raman spectrum of the (0,8)&(0,16)-DWBNNT, which is resulting from the BN stretching along tube axis, including bending deformation of the NBN/BNB bonds along tube axis. Additionally, Obraztsova and coworkers [63] have studied comparative Raman spectra of the multi-walled boron nitride nanotubes (MWBNNTs) samples before and after Al ion modifications have been investigated. Two features in the Raman spectra were observed: one at 1366 cm^{-1} that corresponds to in-plane vibrations between B and N atoms and broad feature around 1293 cm^{-1} in the spectrum of the Al-modified MWBNNTs. The broad peak around 1293 cm^{-1} is consistent with the calculated Raman feature around 1250 cm^{-1} in the spectra of the DW- and SW-BNNTs.

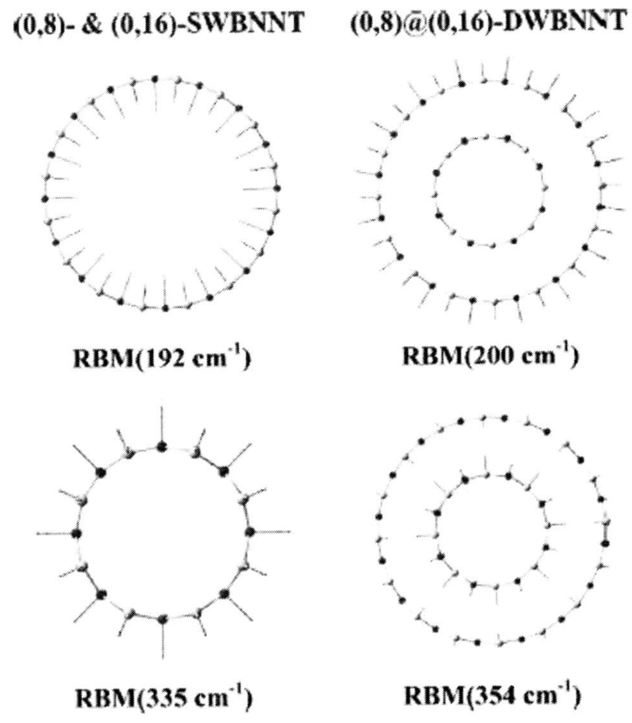

Figure 8. Calculated molecular motions for some vibrational bands of the (0,8)&(0,16)-DWBNNTs and (0,8)- and (0,16)-SWBNNTs.

3.3. Ir Spectra Of Single-Walled And Double-Walled Boron Nitride Nanotube

Zigzag-SWBNNTs: Figure 9A provides calculated IR spectra for the (n,0)-SWNTs, where n ranges from 6 to 19. As evidenced in Figure 9, the calculated IR spectra exhibited seven peaks of symmetries E_1u and A_{1u} are slightly depend on the SWBNNTs diameter. In the range of 1000 to 1550 cm^{-1}, relatively very weak six IR features of symmetries E_{1u} are centered: ~ 1475 ± 25, ~1330 ± 30, ~1230 ± 30, ~1030 ± 5 cm^{-1}, and other two weak peaks with symmetry A_{1u} are centered ~1495 ± 15 and ~1350 ± 15 cm^{-1}. The strongest one with symmetry E_{1u} is centered 1395 ± 30 cm^{-1}. In the range of mid frequency, the calculated IR spectra of the (0,n)-SWBNNTs (n= 6 to 19) exhibited only one weak peak centered 805 ± 15 cm^{-1}. The analytical expressions for this calculated high frequency band as functions of third order in inverse of the (0,n)-SWBNNTs diameter are given by the equations: $\omega(cm^{-1}) = A + \frac{B}{d_t(nm)} + \frac{C}{[d_t(nm)]^2} + \frac{D}{[d_t(nm)]^3}$, where the parameters A(in cm^{-1}), B(in $cm^{-1}.nm$), C(in $cm^{-1}.nm^2$) and D(in $cm^{-1}.nm^3$) are respectively obtained such as: 1508.9, 10.3, -15.6, and 2.4 for the peak (A_{1u}) centered 1495 ± 15 cm^{-1}; 1515.7, -14.9, -2.4, and -3.6 for the peak (E_{1u}) centered 1475 ± 25 cm^{-1}; 1344.7, 130.0, -54.1, and -2.4 for the peak (E_{1u}) centered 1395 ± 30 cm^{-1}; 1357.2, 33.7, -41.0, and 10.2 for the peak (A_{1u}) centered 1350 ± 15 cm^{-1}; 1367.0, -4.9, -17.4, and 0.8 for the peak (E_{1u}) centered 1330 ± 30 cm^{-1}; 1264, 23.2, -37.3, and 3.8 for the peak (E_{1u}) centered 1230 ± 30 cm^{-1}; 1030.5, 28.6, -31.4, and 9.1 for the peak (E_{1u}) centered 1030 ± 5 cm^{-1}; and 824.8, 7.5, -22.3, and 4.1 for the peak (E_{1u}) centered 805 ± 15 cm^{-1}. The plots of the calculated IR features vs. inverse of the tube diameter are given in Figure 9B. In the low frequency region, the IR spectra exhibited many IR features; however, their intensities are extremely weak or vanish as seen in Figure 9A. Furthermore, we provided the vibrational mode assignments and frequencies for the IR spectra of the isolated zigzag-SWBNNTs in Tables 2.

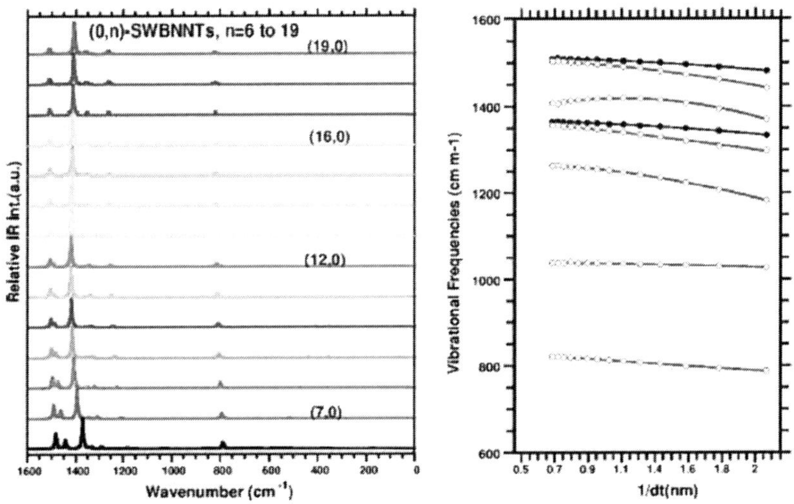

Figure 9. (A) calculated IR spectra of the (0,n)-SWBNNTs, n = 6–19 and (B) the plots of the frequencies of vibrational modes versus $1/d_t$.

Table 2. DFT-calculated IR vibrational frequencies (in cm^{-1}) and assignments for (0,n)-SWBNNT and (0,n)&(0,2n)-DWBNNTs at the B3LYP/6-31G level.

	(0,6)	(0,7)	(0,8)	(0,9)	(0,10)	(0,11)	(0,12)	(0,13)	(0,14)	(0,15)	(0,16)	(0,17)	(0,18)	(0,19)	(0,6)&(0,12)	(0,7)&(0,14)	(0,8)&(0,16)	(0,9)&(0,18)
E_{1u}	788	793	798	804	807	811	813	815	817	818	819	820	821	821	764 813	775 820	786 823	795 824
	Out-of-surface bending deformation of the NBN/BNB bonds on the tube																	
E_{1u}	1182	1207	1224	1235	1242	1248	1252	1255	1258	1260	1261	1262	1264	1264	1253	1197	1238	1257
	Asymmetric stretching vibrations of the NBN/BNB bonds due to the motions of the N and B atoms along circumference direction.																	
E_{1u}	1298	1308	1320	1329	1335	1340	1344	1347	1349	1351	1353	1354	1356	1356	1366	1351	1373	1376
	Bending deformation of the NBN/BNB bonds, including relatively weak BN bond stretching																	
A_{1u}	1333	1343	1349	1353	1355	1358	1359	1361	1362	1363	1363	1364	1364	1365	1332	1332	1347	1363
	BN stretching and bending deformation of the NBN/BNB bonds along tube axis																	
E_{1u}	1370	1394	1407	1414	1418	1419	1419	1418	1416	1414	1412	1410	1408	1406	1439	1433	1434	1429
	Asymmetric stretching vibrations and bending deformations of the BNB/NBN bonds.																	
E_{1u}	1442	1462	1473	1481	1486	1491	1493	1496	1498	1499	1501	1502	1503	1504	1488	1466	1474	1476
	BN stretching, including bending deformations of NBN/BNB bonds																	
$A1_u$	1481	1490	1496	1500	1502	1504	1505	1507	1507	1508	1509	1509	1509	1510	1511	1517	1517	1519
	BN stretching along tube axis, including bending deformations of NBN/BNB bonds																	

DWBNNTs: While Figure 10 provides the calculated IR spectra for the (0,n)&(0,2n)-DWBNNTs, with n ranging from 6 to 9; Figure 11 provides calculated IR spectra of the (0,8)&(0,16)-DWBNNTs and isolated (0,8)- and (0,16)-SWBNNTs for comparison. The calculated spectra of the DWBNNTs 1517 (A_{1u}), 1474 (E_{1u}), 1434 (E_{1u}), 1373 (E_{1u}), 1347 (A_{1u}), 1238 (E_{1u}), 823 (E_{1u}) and 786 (E_{1u}) cm^{-1}, which are correspond the IR features at 1496, 1473, 1407, 1320, 1349, 1224, and 798 cm^{-1} in the spectrum of the (0,8)-SWBNNT; these are calculated at 1509, 1501, 1412, 1353, 1363, 1261, and 819 cm^{-1} in the spectrum of the (0,16)-SWBNNT, as seen in Table 2. Moreover, Y. Bando [62] have studied FTIR spectra of the multi-walled boron (natural ^{11}B and isotope ^{10}B) nitride nanotubes (MWBNNT and $MW^{10}BNNT$). Their FTIR spectra of the MWBNNT and $MW^{10}BNNT$ revealed blue degraded strong IR peak at 1376 and 1392 cm^{-1}, respectively, which is assigned to a B-N stretching deformation vibration mode. This measured IR peak is in good agreement with our calculated IR peaks at 1367 cm^{-1} (E_{1u}, resulting from the bending deformation of the NBN/BNB bonds along tube axis) and 1424 cm^{-1}(E_{1u}, due to the BN stretching along tube axis, including bending deformations of NBN/BNB bonds). Author also observed a relatively weak and broad IR features at ~800 cm^{-1} and suggested that this IR peak is due to the existence of some B-O bonds in their BN nanotubes, see Figure 4 in Ref. [62]. However, our calculated IR spectra of the SWBNNTs and DWBNNTs exhibited IR feature with relatively weak around 800 cm^{-1} is as a result of the out-of surface bending deformation of NBN/BNB bonds on the boron nitride nanotube. Therefore, we suggest that this IR peak (~800 cm^{-1}) may originate from the boron nitride nanotube.

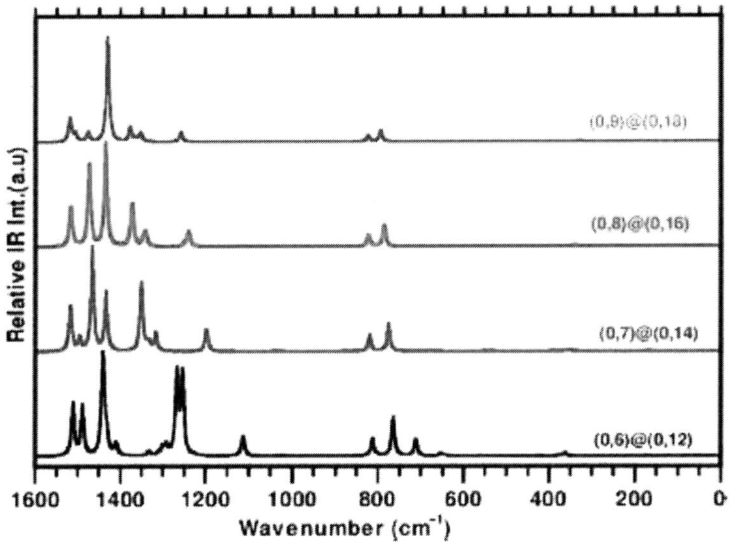

Figure 10. Calculated IR spectra of the (0,n)&(0,2n)-DWBNNT, n = 6–9.

3.4. Electronic Transition Energies Of Dwbnnt And Swbnnts

As mentioned in the introduction to this section, boron nitride nanotubes (BNNTs) can be viewed as modified CNT, but their electronic properties differ from carbon nanotubes. For instance, depending on their chirality and the radius, although carbon nanotubes can be either metallic or semiconducting, all boron nitride nanotubes (BNNTs) are semiconducting materials with a large band. And since the band gap is large, the gap energy is only weakly dependent on the diameter, chirality, and the number of the walls of the tube. Furthermore, owing to their semiconducting character, BNNTs, like CNTs, themselves are also very interesting materials for application in nanoscale devices, and have been considered alternatives to CNTs. The DWBNNTs as well as the doped BNNTs nanotubes may show a dramatic change relative to the isolated nanotube. On account of the strong interactions between electrons and holes in DWBNNTs, the excitonic effects in BNNTs is expected to be more important than in CNTs, since bright (dipole allowed) and dark (dipole forbidden) excitons in DWBNNTs can exhibit qualitatively different optical response. Therefore, the time-dependent DFT (i.e., TD-DFT) method has been applied to investigate the dark transient structures involved in radiationless processes for the DWBNNTs. In this section, we provide the calculated vertical electronic transitions of (0,6)&(0,12)-DWBNNT and (0,6)- and (0,12)-SWBNNTs using DFT and discuss these results in terms of IC and ISC processes.

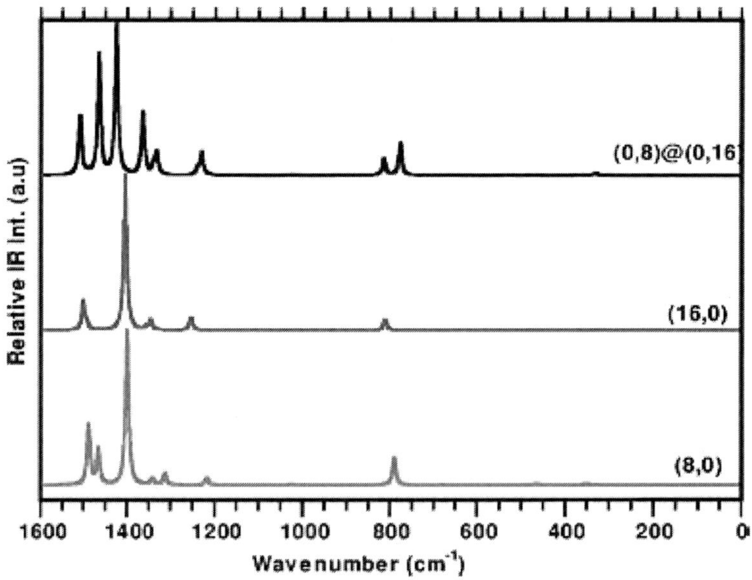

Figure 11. Calculated IR spectra of the (0,8)&(0,16)-DWBNNT and (0,8)- and (0,16)-SWBNNTs for comparison.

The calculated vertical electronic transitions of (0,6)&(0,12)-DWBNNT and (0,6)- and (0,12)-SWBNNTs, as seen in Figure 12 and Table 3, indicated that the lowest electronic energy level (dipole forbidden) of the DWBNNTs are lower as much as about 0.4 eV relative to the (0,6)-SWBNNT and 1.5 eV relative to the (0,12)-SWBNNT. However, when we compare the lowest dipole allowed electronic transitions, the lowest dipole allowed electronic transitions of the DWBNNT are about 1.07 eV and 0.99 eV lower than that for (0,6)- and (0,12)-SWBNNTs, respectively.

Table 3. The calculated vertical electronic transitions, singlet-singlet ($S_0 \rightarrow S_n$ and triplet-triplet ($T_1 \rightarrow T_n$), of the (12,0)&(6,0)-DWBNNT and (12,0)- and (6,0)-SWBNNTs for comparison at the B3LYP/6-31G level of using DFT. Note that the SCF corrected triplet-triplet electronic transitions were calculated as the deference between the calculated global energies of the singlet and triplet sates added to triplet-triplet electronic transitions in order to comparing with the singlet-singlet transitions and where the letters S_0, T_1 and f are respectively the lowest energy level of the singlet, triplet states and oscillator strength.

(0,6)&(0,12)-DWBNNT							(0,12)-SWBNNT			(0,6)-SWBNNT		
$S_0 \rightarrow S_n$				T1→Tn (SCF Corrected)			$S_0 \rightarrow S_n$			$S_0 \rightarrow S_n$		
Exc. St.#	Sym.	eV	f	SYM.	eV	f	Sym.	eV	f	Sym.	eV	f
1:	A"	4.35		A"	4.24		E_1	5.83		A"	4.72	
2:	A'	4.47		A'	4.59	0.0071	E_1	5.95		A'	4.86	
3:	A"	4.67		A"	4.59	0.0072	E_1	5.95		A"	4.86	
4:	A'	4.67		A"	4.93	0.0006	E_1	5.97	0.0871	A'	4.86	
5:	A'	4.89		A'	4.93	0.0007	E_1	5.97	0.0871	A"	5.71	
6:	A"	4.89		A'	4.96		E_1	6.01	0.0239	A'	5.71	
7:	A"	4.90	0.0334	A"	4.96		E_2	6.19		A'	5.83	
8:	A'	4.91	0.0331	A"	4.98		E_2	6.19		A'	5.89	0.0001
9:	A'	4.93		A'	5.01		E_1	6.30	0.8777	A"	5.89	0.0129
10:	A"	4.94	0.0003	A"	5.24		E_1	6.30	0.8777	A"	5.90	0.0316
11:	A'	5.12	0.0055	A'	5.24		A_1	6.36	0.0256	A'	5.90	0.0443
12:	A"	5.12	0.0058	A'	5.27	0.0088	E_1	6.39	0.0168	A"	5.94	0.0002

13:	A"	5.21	A"	5.27	0.0091	E_1	6.39	0.0168	A"	6.05	
14:	A'	5.21	A'	5.44	0.0251	A_2	6.43		A'	6.11	0.0062
15:	A"	5.25	A"	5.44	0.0234	E_2	6.51	0.5031	A"	6.11	0.0063
16:	A'	5.25	A'	5.52	0.0022	E_2	6.51	0.5031	A'	6.20	0.0001
17:	A"	5.26	A'	5.56		E_2	6.52		A"	6.20	0.0001
18:	A"	5.30	A"	5.65		E_2	6.52		A'	6.26	0.0092
19:	A'	5.37	A"	5.67		E_1	6.60	0.0012	A'	6.38	
20:	A"	5.37	A'	5.68		E_1	6.60	0.0012	A"	6.38	

The predicted dipole allowed electronc transitions, S_0(A')→S_7(A") (4.90 eV) and S_0(A')→S_8(A') (4.91 eV), respectively,are due to the HOMO-4(A")→LUMO(A') and HOMO-5(A')→LUMO(A') transitions; S_0(A')→S_{10}(A") (4.94 eV) is as a result of HOMO-6(A")→LUMO(A') transition; S_0(A')→S_{11}(A") (5.12 eV) is mainly due to HOMO(A")→ LUMO+4(A") and HOMO-3(A')→LUMO+2(A')transitions and S_0(A')→ S_{12}(A") (5.12 eV) is mainly because of HOMO-3(A')→LUMO+1(A") and HOMO(A")→LUMO+5(A') transitions. These calculated transitions, together with the plotted electron densities in the HOMOs and LUMOs, as seen in Figure 4B,indicated that first three of five dipole allowed electronic transitions of the (0,6)&(0,12)-DWBNNT, S_0(A')→S_7(A")/ S_8(A')/ S_{10}(A'), originating from the electron transfer from the outer-shell to the inner-shell. These results are clear evidence of the charge transfer from the other shell to the inner shell. The dipole allowed electronic transitions S_0(A')→ S_{11}(A') shows the electron excited from both inner- and other-shells to mostly inner shells, also there is a significant sigma-bonding interactions between inner- and outher-shells. Finally, the S_0(A')→ S_{11}(A') transition indicate that the transitions from both shells to the excited state mainly are due to sigma-bonding interactions. We also calculated the triplet-triplet transitions, which produce many dipole allowed transitions. The SCF corrected electronic transitions of the singlet-singlet and triplet-trpilet of the (0,6)&(0,12)-DWBNNT, together with the singlet-singlet transitions, are given in Figure 12. As seen in Figure 12 and Table 3, upon irradiation, there is the possibility of a system that can undergo internal conversion (IC) and intersystem crossing (ISC) processes via vibroelectronic coupling, besides the photochemical and other photophysical processes. The IC and ISC processes would able to be expected when taking account of the small distance between the electronic energy levels and range of the vibrational spectra of the DWBNNTs.

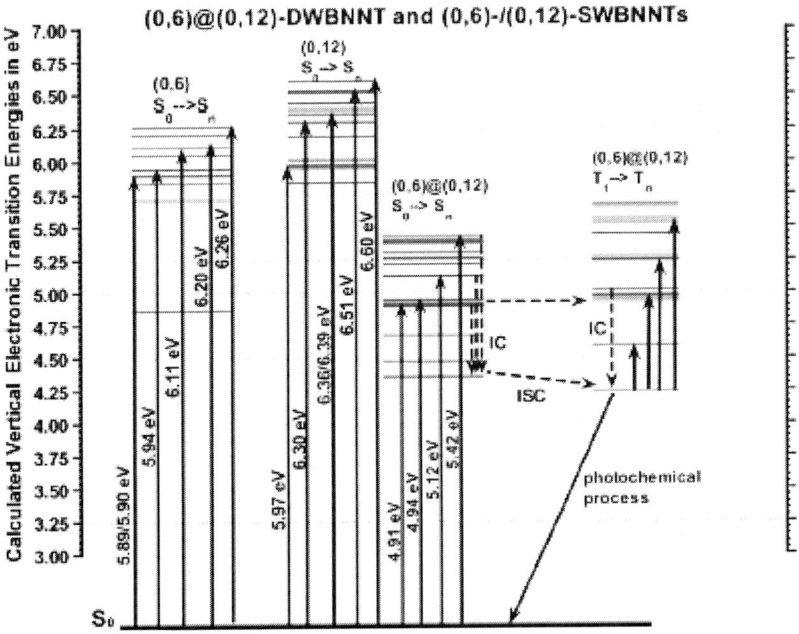

Figure 12. Calculated vertical electronic transitions, singlet–singlet (S0→Sn and triplet–triplet (T1→Tn) for (0,6)&(0,12)-DWBNNT and (0,6)- and (0,12)-SWBNNTs. The vertically solid arrow indicated dipole allowed transitions. The broken-arrows display possible internal conversion (IC) and intersystem crossing (ISC) processes.

C. H. Lee and coworkers [64] have measured absorption spectrum of the suspension of BNNTs in ethanol by using UV–visible absorption spectroscopy (HP 8453 Spectrophotometer). The authors observed three absorption bands at ~5.9 eV (very strong) and ~4.78 eV (weak), and ~3.7 eV (very weak) in the UV-visible spectrum and suggested that the band at about 4.75 eV originates from the intrinsic dark exciton absorption band; the relatively small band at ~ 3.7 eV was due to the defects of the boron nitride nanotubes (BNNTs), and the stronger band at 5.9 eV was as results of the optical band gap of BNNTs. For the (0,6)&(0,12)-DWBNNT, as seen in Table 3, our calculated electronic transitions produced a few dipole allowed electronic transitions below 5.37 eV such as: $S_0 \rightarrow S_7/S_8$ at 4.90 eV (with the f = 0.0334), $S_0 \rightarrow S_{10}$ at 4.94 eV (f = 0.0003), $S_0 \rightarrow S_{11}/S_{12}$ at 5.25 eV (f = 0.0055), which are in good agreement with this measured band at about 4.78 eV.

Furthermore, for the (0,6)- and (0,12)-SWBNNTs, the calculations exhibited the lowest dipole allowed electronic transition around 5.9 eV, which is in accordance with the measured strong optical band at 5.9 eV. The lowest dipole forbidden transitions are predicted at 4.35, 4.72, and 5.83 eV for the (0,6)&(0,12)-DWBNNT, (0,6)- and (0,12)-SWBNNTs, respectively. Consequently, this experimentally measured UV-visible spectrum might be an evidence for the formation of the (0,6)&(0,12)-DWBNNT, the observed absorption band (at ~4.78 eV) may due to the $S_0 \rightarrow S_7/S_8$ (4.90 eV), not due to the intrinsic dark exciton as suggested by authors.

Furthermore, Figure 4A provides the calculated electron density of (0,6)&(0,12)-DWCNT (double-walled carbon nanotube), showing that the first four highest occupied molecular orbitals (from HOMO to HOMO-3 with the A_{1u}, A_{2g} and $2E_{1g}$ symmetries, respectively) belong to the outer-shell, and the next highest occupied molecular orbitals from HOMO-4 to HOMO-24 include both inner- and outer-shells of (0,6)&(0,12)-DWCNT. The lowest unoccupied molecular orbital LUMO (E_{1u}), lying about 0.780 eV above the HOMO (A_{1u}), belongs to the outer-shell, while the next one (B_{2u}) belongs to the inner-shell and lies 0.849 eV above the HOMO (A_{1u}). The calculated electron density also indicates that an intratube (inner and outer tube) interaction may possibly take place in the excited state: the LUMO+7 with A_{2u} symmetry and 2.494 eV above the HOMO (A_{1u}), LUMO+8 (E_{1u}; 2.557 eV), LUMO+10 (E_{1g}; 2.563 eV) and LUMO+15 (E_{1g}; 3.637 eV). The intratube CC σ-bonding interaction in the excited state may lead to an intertube charge transfer, which can be observed by a significant change in the tangential modes (TMs) of Raman spectra when the tube is excited to its intratube charge transfer state. The TM may provide information not only about the metallic or semiconducting character of nanotubes, but also on the inner-outer tube (intratube) charge transfer.

Similarly, the calculated vertical electronic transitions for the (0,n)&(0,2n)-DWBNNT and (0,n)- and (0,2n)-SWBNNTs, n= 8 and 9, at the same level of the theory. The calculated singlet-singlet ($S_0 \rightarrow S_n$and triplet-triplet ($T_1 \rightarrow T_n$) electronic transitions are given in Table 4.

For the (0,8)&(0,16)-DWBNNT, the predicted dipole allowed electronc transitions, $S_0 \rightarrow S_2$/ S_3 (5.39 eV, mainly due to the H-1 →L and H→L) and $S_0 \rightarrow S_6$ (5. 47 eV, mainly due to the H-3 →L). These calculated transitions, in conjunction with the plotted electron densities in the HOMOs and LUMOs, as seen in Figure 4C, indicated that first three dipole allowed electronic transitions of the (0,8)&(0,16)-DWBNNT, $S_0 \rightarrow S_{2/3}$/ S_6 originating from the electron transfer from the outer-shell to the inner-shell. These results of the calculations provide not only clear evidence for the charge transfer from the other shell to the inner shell, but also there is a significant BB σ-bonding interaction between the inner- and outer-shells. As seen in Table 4, the lowest dipole allowed vertical electronic transition of the (0,8)&(0,16)-DWBNNT ($S_0 \rightarrow S_2$; 5.39 eV) lies 0.61 and 0.52 eV below the lowest allowed transitions of the $S_0 \rightarrow S_6$ and $S_0 \rightarrow S_4$ for the (0,8)- and (0,16)-SWBNNTs, respectively.

Furthermore, upon irradiation, a system can undergo internal conversion (IC) and intersystem crossing (ISC) processes, besides the photochemical and other photophysical processes. Transient intermediates are likely to form in the IC and ISC radiationless processes, which is also known as "dark processes". Our calculations also indicated that possibilities of the IC and ISC processes via vibroelectronic coupling, besides the photochemical and other photophysical processes. For instance, based on the calculated electronic transitions as seen in the Table 4, when the (0,8)&(0,16)-DWBNNTs are excited, all of the excited nanotubes may not directly return back to their ground state by emission of a photon, $S_{k>0} \rightarrow S_0$ transition, but some of them may return back to their ground states (S_0) by the IC (internal conversion), for instance, when the system is excited into a higher vibroelectronic state (S_6, 5.47 eV), it may undergo into the S_1 state (5.39 eV) via vibrational coupling between these two states before undergoing additional vibrational relaxation back to the lowest singlet electronic energy level (S_1), which is called internal conversion (IC), then, followed by transition from the second lowest singlet electronic energy level S_1(5.39 eV) to S_0 by emission of a photon is so-called fluorescence. An alternate pathway for a molecule in the S_1 state involves an intersystem crossing (ISC) by the nanotube into the lowest triplet electronic state T_1 (5.28 eV). From T_1, the nanotube can undergo radiative de-excitation via a much slower process, which is known as phosphorescence ($T_1 \rightarrow S_0$ transition) such as illustrated in Figure 12.

Likewise, for the (0,9)&(0,18)-DWBNNT, the calculations indicated that the lowest dipole allowed transition ($S_0 \rightarrow S_3$, 5.69 eV) lies 0. 18 and 0.21 eV below the lowest allowed transitions of the (0,9)- and (0,18)-SWBNNT. Additionally, as seen in Table 4, the calculations also indicated that the possibilities of the IC from the S_k (k=3,4,7-9, and 14) to S_1 as well as ISC proces from the singlet electronic state S_1(5.67 eV) to T_1 (5.71 eV) for the (0,9)&(0,18)-DWBNNT. The calculated dipole allowed vertical electronic transitions may be summarized as following: the transition $S_0 \rightarrow S_{3/4}$ (5.69 eV and f=0.1656) is predominantly as result of the electron excitation mostly from the outer shell to the inner shell (H→L+1, H-1→L, H-6→L+1/2), including excitations from inner shell to the outer (H-2→L+1/2 and H-3→L+1/2); $S_0 \rightarrow S_7$ (5.73 eV and f=0.0060 is mainly as result of the electronic excitation from the outer shell to the inner shell (H→L+1, H-1→L+2, H-6→L), including relatively weak contribution from inner shell to the outer (H-3→L+5 and H-2→L+6); and the transitions $S_0 \rightarrow S_{8/9}$(5.74 eV and f=0.0007) and $S_0 \rightarrow S_{14}$ (5.78 eV and f=0.0077) are as result of the electronic excitation from the outer shell to the outer shell (H-5→L+3/4 and H-4→L+3/4), as shown in Figure 4 D.

The key conclusions on the calculated electronic spectra indicates that the first dipole allowed electronic transitions of the (0,n)&(0,2n)-DWBNNTs (n = 6,8, 9) lead to a charge transfer process from outer shell to the inner shell. Moreover, there is a significant intertube σ-bonding interactions between the inner- and outer-shells occurs with decreasing distance between the interwall of the DWBNNTs, in contrast, for the (0,9)&(0,18)-DWBNNT, there is a relatively weak contributions to the charge transfer process from the inner-shell to outer-shell.

4. COVALENTLY FUNCTIONALIZED ZIGZAG-SWCNTS

Carbon nanotubes have broad range of potential applications from medical to industry fields due to their unique structural, mechanical, and electronic properties, as mentioned in the introduction section. Different functionalization methods such as chopping, oxidation, wrapping and irradiation of the CNTs can lead to active bonding sites on the surface of the nanotubes.

In this section, we calculate, for covalently functionalized carbon nanotubes (f-CNTs), such parameters as the curvature energies referenced, IR and Raman spectra, and vertical electronic transitions. The latter one may be important to understand the optical mechanism for the charge transfer between functional group(s) and CNT as well as internal conversion and intersystem crossing, as well photochemical process that may occur.

The structure of the functionalized-single-walled carbon nanotubes, f-(n,0)-SWCNTs, constructed of functional group(s) covalently bound on the (n,0)-SWCNTs,of two unit cell length, has been investigated. The most stable of the geometry has been obtained by full optimization without any symmetry restriction. The optimized structure indicated that the cylindrical shape of the nanotube is altered to an elliptical form when two molecules attached to the surface of CNT; but the structure remains almost cylindrical with C_4 symmetry, when four functional groups are bound. When we used benzenesulfonic acid (ph-SO_3H; $C_6H_5SO_3H$) as a functional group that covalently bonds on the surface of the (n,0)-SWCNTs, n=6 to 12, the curvature energy per hexagon, (ΔE[f–(n,0)–SWCNTs]ΔEf-n,0-SWCNTs),of the functionalized-(n,0)-SWCNT calculated relative to that of the corresponding isolated species is given by the following equation:

$$\Delta E[f-(n,0)-SWCNTs; in\ eV] = \frac{E[f-(n,0)-SWCNT]-E[f]}{2n} - \frac{E[(n,0)-SWCNTs]}{2n} \quad (9)$$

where E[f-(n,0)-SWCNTs], E[f] and E[(n,0)-SWCNTs] indicates the global energy of functionalized-(n,0)-SWCNT, isolated benzenesulfonic acid ($C_6H_5SO_3H$) and isolated (n,0)-SWCNT, respectively,The f and n stand for the functional group and chiral index of the zigzag-CNTs. The plot of the calculated relative curvature energy is given in Figure 13. As seen in the Figure 11, the relative curvature energy for the metallic and semiconducting CNTs are well separated. Based on the predicted value of the energies, the results suggested that the covalently functionalization of the SWCNT, with small diameters, are energetically more stable than that with large diameters for the metallic nanotubes. However, for semiconducting nanotubes, the functionalization of the tube is favorable, but the functionalization of the (11,0)-SWCNT is more favorable than (10,0)-SWCNTs. In order to make a correct overall assessment, we need to more data, at least for semiconducting zig-zag nanotubes.

4.1. Raman Spectra Of Functionalized Zigzag-Swcnts

The calculated nonresonance Raman spectra for the covalently functionalized-(n,0)-SWCNTs with benzenesulfonic acid (-ph-SO3H) and the isolated (n,0)-SWCNTs (where n = 7 to 10), as well as the spectrum of the functionalized (7,0)-SWCNT with the carboxylic acid (-COOH), for comparison, are shown in Figure 14. Because of the similarity of the Raman spectra of the f-SWCNTs, here we only discuss the Raman spectra for the functionalization of the (7,0)-SWCNT with the benzenesulfonic acid and carboxylic acid, and the spectrum of the isolated (7,0)-SWCNT. The Raman spectra of both functionalized (7,0)-

SWCNT exhibited many new features relating to the spectrum of the isolated (7,0)-SWCNT as well as shift in the peak positions. The predicted results are summarized below.

Table 4. The calculated vertical electronic transitions, singlet-singlet ($S_0 \rightarrow S_n$ and triplet-triplet ($T_1 \rightarrow T_n$), of the (0,8)&(0,16)-DWBNNT and (12,0)- and (6,0)-SWBNNTs for comparison at the B3LYP/6-31G level of using DFT. Note that the SCF corrected triplet-triplet electronic transitions were calculated as the deference between the calculated global energies of the singlet and triplet sates added to triplet-triplet electronic transitions in order to comparing with the singlet-singlet transitions and where S_0 and T_1 is respectively the lowest energy level of the singlet and triplet states.

	(0,8)&(0,16)-DWBNNT				(0,8)-SWBNNT		(0,16)-SWBNNT		(0,9)&(0,18)-DWBNNT				(0,9)-SWBNNT		(0,18)-SWBNNT	
	$S_0 \rightarrow S_n$		$T_1 \rightarrow T_n$		$S_0 \rightarrow S_n$		$S_0 \rightarrow S_n$		$S_0 \rightarrow S_n$		$T_1 \rightarrow T_n$		$S_0 \rightarrow S_n$		$S_0 \rightarrow S_n$	
n	eV	f	eV	f	eV	f	eV	f	eV	f	eV	f	eV	f	eV	f
1	5.39		5.28		5.61		5.79		5.67		5.71		5.77		5.79	
2	5.39	0.1039	5.32	0.0003	5.61		5.88		5.68		5.74	0.0002	5.86		5.87	
3	5.39	0.1039	5.32	0.0003	5.69		5.88		5.69	0.1656	5.77	0.0004	5.86		5.87	
4	5.46		5.37		5.80		5.91	0.1868	5.69	0.1652	5.78		5.87	0.0215	5.90	0.2880
5	5.46		5.37		5.87		5.91	0.1868	5.70		5.85	0.0021	5.87	0.0215	5.90	0.2880
6	5.47	0.0007	5.49		6.00	0.0255	5.94	0.0268	5.70		5.92	0.0002	5.95	0.0135	5.91	
7	5.47		5.50		6.00	0.0255	5.96		5.73	0.0060	5.94	0.0001	6.06		5.91	
8	5.47		5.50		6.02		5.96		5.74	0.0007	5.95	0.0058	6.06		5.93	0.0287
9	5.52		5.67		6.02		6.16	2.0492	5.74	0.0007	6.00	0.0018	6.06		6.06	
10	5.67		5.82	0.0594	6.04		6.16	2.0492	5.76		6.01	0.0038	6.06		6.06	
11	5.67		5.82	0.0594	6.07	0.0164	6.17		5.76		6.07	0.0021	6.09		6.11	2.5027
12	5.67		5.92	0.0167	6.11		6.17		5.78		6.15	0.0077	6.09		6.11	2.5026
13	5.69		5.92	0.0167	6.11		6.27	0.0432	5.78		6.16	0.0049	6.16		6.24	0.0165
14	5.69		5.94		6.14		6.27	0.0432	5.78	0.0077	6.23	0.0151	6.16		6.24	0.0165
15	5.74		6.00		6.14		6.31		5.83		6.24	0.0103	6.28		6.28	
16	5.74		6.00		6.16		6.31		5.83		6.26	0.0215	6.28		6.28	
17			6.03		6.17		6.34	0.0211			6.28	0.0065	6.30	0.0918	6.34	
18					6.18		6.35	0.0587					6.30	0.0918	6.34	
19					6.33	0.2616	6.35	0.0587							6.35	0.0334

In the low energy region below 600 cm^{-1}, 1) one of the important Raman peak, which is the radial breathing mode (RBM), was predicted at 410 cm^{-1} in the isolated (7,0)-SWCNT shifted not only to 390 and 385 cm^{-1} in the spectra of the (7,0)-SWCNT functionalizes with benzenesulfonic acid and carboxylic acid, respectively, but also enhanced in both spectra; 2) the relatively peaks at 109 and 111 cm^{-1} result from the elliptical deformation of the carbon nanotube are respectively shifted to 75 and 121 cm^{-1} (in the spectrum of the (7,0)-SWCNT functionalizes with benzenesulfonic acid), and to 95 and 134 cm^{-1} in the Raman spectrum of the functionalization of the (7,0)-SWCNT with carboxylic acid, the intensity enhanced in both spectra of the functionalized tube; 3) a doubly degenerated peak predicted at 284 cm^{-1} (as a result of diagonal expansion of the tube) in the Raman spectrum of the isolated tube is split into well separated two peaks and appeared at about 250 and 306 cm^{-1} in the spectrum of each (7,0)-SWCNT functionalizes with benzenesulfonic acid and carboxylic acid; 4) a relatively very weak peak at 500 cm^{-1} in the spectrum of the isolated tube appeared at same position, but its intensity significantly enhanced, in the calculated both Raman spectra of the (7,0)-SWCNT functionalizes with benzenesulfonic acid and carboxylic acid, which is ; 5) many relatively weak Raman features (result from the out-of-plane structural deformation of the functional groups) appeared below 600 cm^{-1} as seen in the Figure 14 and 15. In the range from 600 to 1250 cm^{-1}, the Raman spectra of the f-(n,0)-SWCNT exhibited many relatively medium, weak and very weak new Raman peaks beside the peaks appeared at 760, 794 and 911 (very weak) cm^{-1} in the Raman spectra of the f-(n,0)-SWCNT. For instance, in the Raman spectrum of the (7,0)-SWCNT functionalizes with benzenesulfonic acid, the peaks with relatively intense at 1120 cm^{-1} (due to the structural deformation of the tube, including wagging of CH bonds of the benzene ring); at 1138 cm^{-1} (as a result of asymmetric CSO bond straching and OH bond wagging, including relatively weak bending deformation of the benzene ring); at 1142 cm^{-1} (structural deformation of the tube due to the CC streching, accompanied by wagging of Hs on the benzene ring), and the peak at 1185 cm^{-1} is owing to asymmetric CSO bond stretching and wagging of OH bond. The Raman spectrum of the (7,0)-SWCNT functionalizes with carboxylic acid exhibited relatively strong Raman features at 1122 cm^{-1} (caused by structural deformation of the nanotube, including OH bond wagging); 1146 cm^{-1} (by reason of asymmetric streching of CCO(H) bond, including tube deformation), and the calculated Raman peak at 1181 cm^{-1} is due to asymmetric streching of CCO bonds, including tube deformation. The Raman peaks at 760 cm^{-1} (due to expansion of the tube along the tube axis) and 795 cm^{-1} (as a result of out-of-surface bending deformation of the tube) in the Raman spectrum of the isolated

(7,0)-SWCNT at the same positions of the f-SWCNT). A strong peak at around 1225 cm^{-1} in the spectra of the (7,0)-SWCNT and f-(7,0)-SWCNT is completely originates from the wagging of the CH bond at end of the tube. There are also many very weak Raman features appeared in this range from 600 to 1250 cm^{-1}. In the range from 1300 to 1800 cm^{-1}, two peaks at 1300 cm^{-1} (weak) and 1330 cm^{-1} (strong) in the spectrum of the (7,0)-SWCNT functionalizes with benzenesulfonic acid and with carboxylic acid, as a result of symmetric stretching of CCC bonds and bending deformations along tube axis, which correspond to a relatively weak and doubly degenerate Raman feature at 1305 cm^{-1}. A doubly degenerated peak (relatively very weak) at 411 cm^{-1} (result from asymmetric stretching of CCC bonds within the tube) in the Raman spectrum of the SWCNT is split into two weak peaks at about 1390 and 1405 cm^{-1} in the Raman spectrum of the functionalized (7,0)-SWCNT. The Raman peak with medium intense at 1486 cm^{-1}, resulting from CC bond stretching within the nanotube, corresponds to the peak at ~1504 cm^{-1} in the Raman spectrum of the functionalized (7,0)-SWCNT. The strongest and doubly degenerate Raman peak at 1574 cm^{-1} in the isolated (7,0)-SWCNT, resulting from asymmetric stretching of the CCC bonds along circumference direction of the tube, is blue shifted to nearly degenerated peak at 1590 and 1595 cm^{-1}, as a result of the CC bond stretching within the tube, in the Raman spectra of the f-(7,0)-SWCNT. In this range from 1300 to 1800 cm^{-1}, the Raman spectra of the (7,0)-SWCNT functionalizes with benzenesulfonic acid and with carboxylic acid showed many new Raman features. For example, the strongest peaks appeared at ~1380 and ~1390 cm^{-1} are as a result of asymmetric tube deformation due to the CC bonds stretching, which is not shown in the isolated (7,0)-SWCNT. The peaks at 1373 and 379 cm^{-1} in the spectrum (7,0)-SWCNT functionalizes with benzenesulfonic are mainly due to the asymmetric stretching of the OSO bond and wagging of the OH bond, including asymmetric stretching of the CCC bonds of the benzene ring. The peaks: at 1471 and 1482 cm^{-1}, which is the result of the CC bond stretching within the tube; at , 1548, and 1557 cm^{-1} is due to asymmetric CCC bond stretching within the tube, however, the peak at 1547 cm^{-1} is entirely due to symmetric stretching of the CC bonds of the benzenesulfonic acid . Furthermore, the predicted Raman peak at 1650 cm^{-1} is due to CC bond stretching of the benzene ring, including CH bond wagging on the benzene ring. A very weak peak at 1806 cm^{-1} is as a result of the CO stretching of the carboxylic acid only. As a result of the (7,0)-SWCNT functionalizes with benzenesulfonic acid and with carboxylic acid (f-(7,0)-SWCNT), the key conclusions on these calculated Raman spectra of the f-(7,0)-SWCNT are summarized below: 1) the RBM is red shifted as much as 25 cm^{-1}; 2) many new peaks appeared in the disorder (D) mode range from 1300 to 1450 cm^{-1}, which is due to the structural deformation of the tube and of the functional groups bound to the tube (7,0)-SWCNT); 3) the tangential (or G) mode is blue shifted as much as 20 cm^{-1}, as a result of the functional groups bound to the tube; 4) above the G-mode, appeared new Raman feature in the spectra of the f-(7,0)-SWCNT belong to the functional groups (benzenesulfonic acid and carboxylic acid); 5) the new Raman features are found to appear along the spectrum, which is owing to the combination of the structural deformation of the tube and the functional groups; 6) for the benzenesulfonic acid, while the CH bond stretching mode occurred range from 3200 to 3240 cm^{-1}, the OH bond stretching appear at 3703 cm^{-1}; for the carboxylic acid, the OH bond stretching is predicted at 3678 cm^{-1}; the CH bond stretching of the tube are predicted in the range from 3172 to 3200 cm^{-1}.; 7) the RBMs of frequency in the calculated Raman spectra of the functionalized (n,0)-SWCNT , (n=6 to 11) are slightly red-shifted relative to that for isolated SWCNTs as seen in Figure 15. The relative shift in frequency of the RBM decreases with increasing tube diameter.

It is worth nothing that the relative intensity of the peaks in the resonance Raman spectra significantly change. Because of the technical difficulty and calculation time, it is very difficult to calculate resonance Raman spectra. Furthermore, in the low frequency region below 600 cm^{-1}, there are many relatively very weak Raman peaks, which result from out-of-plane motion, or twisting of the phenyl group. These types of Raman bands of the functionalizated the CNTs may significantly enhanced in the resonance Raman spectrum (RRS) since there is a significant dipole-dipole interaction between the functional groups. This may play a crucial role and might be used as signature for the alignment of the CNTs in two dimensional networks, but also, the presence of additional bands may lead to the erroneous conclusion that more than one type of SWNT is present in the sample. For instance, the Raman band(s) resulting from out-of-plane motions are dramatically enhanced when dye molecule aggregate, and are referred to as J- or H- type aggregates.[48(a-d)]

New Raman peaks appeared around 1550 cm^{-1} due to the symmetric stretching of the CCC bonds and rocking of CH bonds in phenyl group of the benzenesulfonic acid. Several new Raman peaks result from only benzenesulfonic acid or combination of benzenesulfonic acid and nanotube dispersed throughout the spectrum. The Raman peak resulting from the stretching of CC sigma bonding between benzenesulfonic acid and SWCNTs is very weak and appear at about 1208 cm^{-1}. In the low frequency region, there are many relatively very weak Raman peaks below 600 cm^{-1}, which result from out-of-plane motion, or twisting of the phenyl group. These type of Raman bands of the functionalized-CNTs can play a crucial role and might be used as signature for the alignment of the CNTs in two dimensional networks. For instance, the Raman band(s) resulting from out-off plane motions are dramatically enhanced when dye molecule aggregate, and are referred to as J- or H- type aggregates.[48(a-d)] It is also worth that the calculations produced nonresonance Raman spectra which differ from the resonance Raman spectra in terms of intensity. Furthermore, the CH stretching of the end group of the CNT appear at around 3185 cm^{-1}, the CH stretching of the benzenesulfonic acid and OH stretching of the carboxyl group are, respectively, at about 3590 and 3680 cm^{-1}.

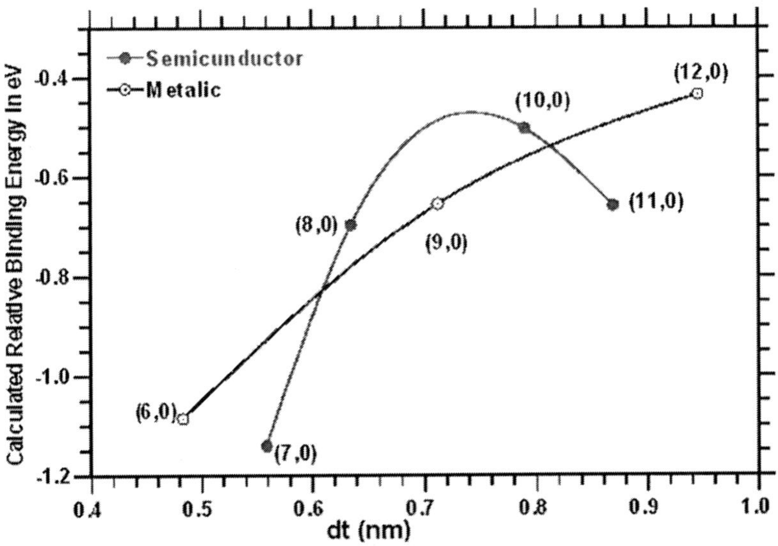

Figure 13. Calculated binding energies of the (n,0)-SWCNTs covalently functionalizes with the benzene sulfonic acid ((n,0)-SWCNTs-ph-SO_3H, n = 6–12). Energetically more stable covalently functionalized (n,0)-SWCNT (f-(n,0)-SWCNTs; n = 6–12) was predicted by using the equation: ΔE[f-(n, 0)-SWCNTs : in eV] = E[f-(n, 0)- SWCNT]/2n – [E(f) + E[(n, 0)-SWCNTs]/2n. Where ΔE[f-(n,0)-SWCNTs] is the energy difference between the total energy of the f-(n,0)-SWCNTs per the number of hexagons in the tube (E[f-(n,0)-SWCNT)/2n]) with reference to the total energy of their corresponding isolated (n,0)-SWCNTs per the number of hexagons in the tube (E[(n,0)-SWCNT)/2n]) and the total energy of the functional groups (E(f)/2n). The letters n and 2n stand for the chiral index of the zigzag-SWCNTs and the number of hexagon in the nanotube, respectively. See Section 3.1 for more detail.

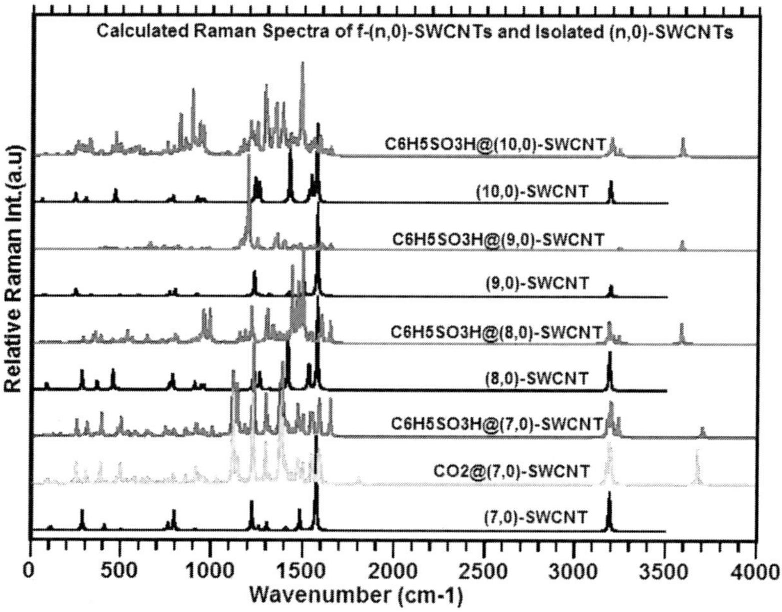

Figure 14. Calculated Raman spectra of the functionalized (n,0)-SWCNTs,benzenesulfonic acid, carboxylic acid, and isolated (n,0)-SWCNTs, n = 7 to 10.

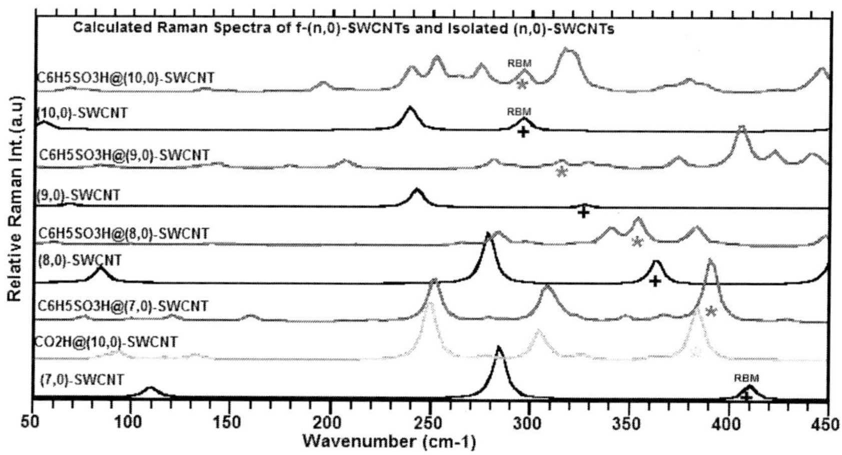

Figure 15. Calculated RBMs of frequencies in Raman spectra of the functionalized (n,0)-SWCNTs with benzenesulfonic acid and carboxylic acid, as well as isolated (n,0)-SWCNTs: n = 7 to 10.

The RBMs of frequency in the calculated Raman spectra of the functionalized SWCNT are slightly red-shifted relative to that for isolated SWCNTs as seen in Figure 15. The relative shift in frequency of the RBM decreases with increasing tube diameter.

4.2. Ir Spectra Of Functionalized Swcnts

As provided in Figure 16, the predicted IR spectra of the (n,0)-SWCNT exhibits strong IR peaks centered at 890 and 845 cm^{-1}; however, the IR spectra of the functionalized (n,0)-SWCNTs display many new strong with relatively weak IR peaks dispersed through spectra, such as at 1650, 1275, 1150, 791, 570, 380, 143 cm^{-1}. Also, in range of 3000-4000 cm^{-1}, the CH and OH stretching modes of the benzenesulfonic acid and carboxylic acid are found to appear at around 1590 and 1670 cm^{-1}, respectively. The C=O bond resulting from C=O stretch of thecarboxyl groups, which is experimentally observed at 1782 cm^{-1} in the FTIR spectra of MWNT, after electron-beam irradiation by Eun-Ju Lee*et al.*[50], is predicted at 1800 cm^{-1} from the calculation.

The peaks found around 1650 cm^{-1} are mainly due to the C-C stretching and CCC bonding deformations; asymmetric and symmetric stretching of the O=S=O group in the benzenesulfonic acid group are found at 1275 and 1150 cm^{-1}, respectively; S-OH stretching appears at 780 cm^{-1}; bending deformation of the SO_3H,mimicking opening and closing of an umbrella,appears at 570 cm^{-1}; out-off plane motion of the phenyl group of the benzenesulfonic acidappears at 380 cm^{-1}; and twisting of the O=S=O bend appears at about 143 cm^{-1}.

4.3. Vertical Electronic Transitions Of Functionalized Swcnts

We calculated the vertical electronic transitions for (n,0)-SWCNTs functionalized with benzenesulfonic acid. The functionalized-SWCNTs were constructed as two- and four- functional groups covalently attached to (7,0)/(9,0) and (12,0)/(8,0)-SWCNTs with length equivalent to two unit. Table 5 provides calculated electronic transitions of functionalized and isolated SWCNTs; selected calculated electron density for the HOMOs and LUMOs states involved in the electronic transitions are provided in Figure 17. The results of the calculations clearly indicate that both of the dipole allowed and forbidden electronic transitions are lowered as much as 0.8 eV relative to the transition energies of thecorresponding isolated SWCNT. Furthermore, the calculations also showed that below 2.5 eV there is no electron transfer from the nanotube to the functional group, or vice versa. However, the calculated electronic densities suggest that there would be intrasystem charge transfer between molecule and the nanotube. Because of the distance among the electronic energy levels is very small for some of the dipole allowed and forbidden electronic transitions, radiationless transitions are expected as a result of vibrational coupling or surface touching of the electronic potential energy surfaces. Coupling maybe very large and might lead to internal conversion (IC), again due to vibroelectronic coupling, which might be observable via fluorescence spectroscopic techniques, as discussed and illustrated in Figure 1in the introduction section. We also would like to point out that while isolated SWCNTs exhibit one or a few dipole allowed electronic transitions below 2.5 eV, the functionalized SWCNTs produced many dipole allowed electronic transitions compared with the corresponding isolated SWCNTs, in addition to lowered electronic transitions.

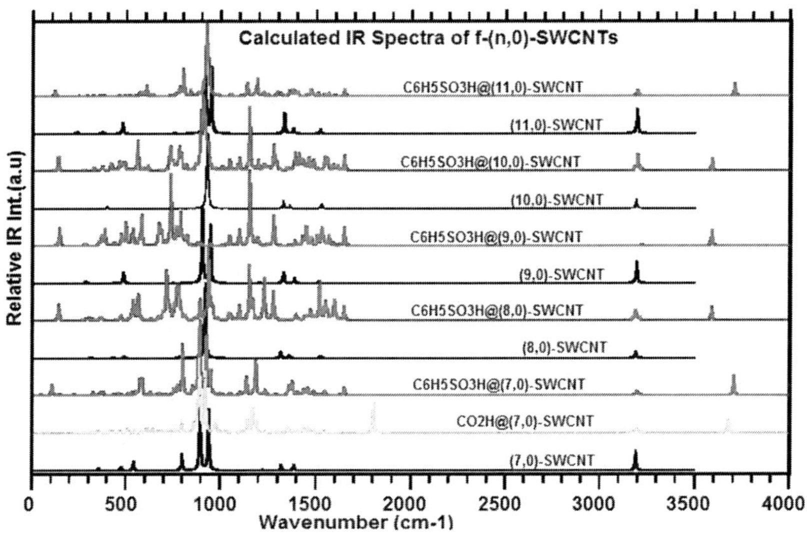

Figure 16. Calculated IR spectra of the (n,0)-SWCNTs functionalized with benzenesulfonic acid, carboxylic acid and isolated (n,0)-SWCNTs: n = 7 to 10.

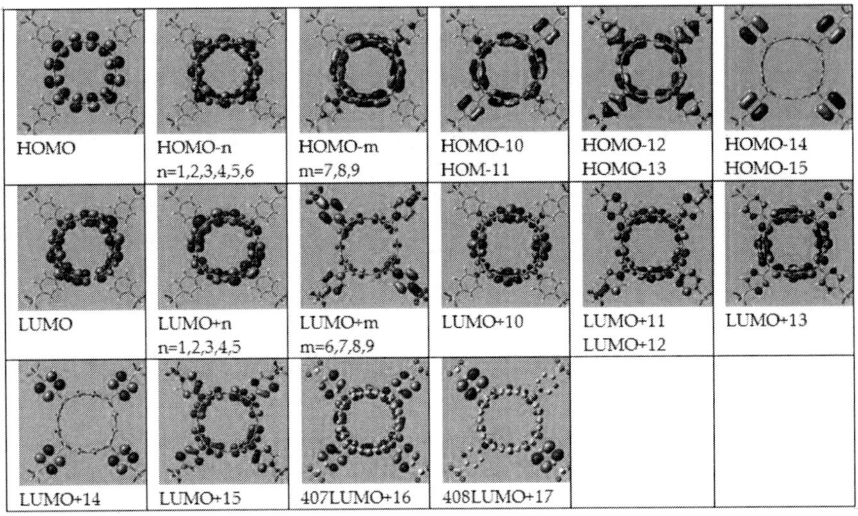

Figure 17. Calculated electron densities in the HOMO and LUMO states for the functionalized (12,0)-SWCNT with benzenesulfonic acid ($C_6H_5SO_3H$).

Calculated vertical electronic transitions, up to 2.53 eV, exhibited many dipole allowed and forbidden electronic transitions. The transitions up to 2.04 eV are due to transitions from the HOMOs of the SWCNT to the LUMO of the SWCNT. Above the 2.04 eV, calculation indicates the existence of charge transfer from the HOMOs of the SWCNT to the LUMOs of the benzenesulfonic acid ($-C_6H_4SO_3H$). For instance, the dipole allowed electronic transitions occur at 2.208, 2.232 and 2.523 eV, as a results of the transitions from the HOMOs of the (12,0)-SWCNT to the molecule only: H→L + 8, H→L + 9 and H-1→L + 9, respectively. As seen in Table 5, there are many dipole allowed electronic transitions from the HOMO of the SWCNT only to the LUMOs of both SWCNT and the benzenesulfonic acid. The results of the calculated vertical electronic transitions of functionalized nanotube ($C_6H_4SO_3H$@(12,0)-SWCNT) indicate that there is a charge transfer mechanism from the nanotube to the molecule above 2.0 eV. The small distance between the electronic transitions suggest that there would be internal conversion (IC) via vibrational coupling as much as 0.43 eVwhen taking account of the spectral range of the vibrational spectrum of the functionalized nanotube. These spectroscopic properties can be observable by quenching of the fluorescence and by Raman and IR spectroscopy. For the ($C_6H_4SO_3H$@(12,0)-SWCNT, the calculated electronic transitions up to 2.41 eV does not indicate any charge transfer process. However, when one examines the calculated energy levels of the HOMOs and the LUMOs it is possible for charge transfer processes to occur above 2.41 eV.

Table 5. Calculated vertical electronic transition energies (Te; in eV), $S_0 \rightarrow S_k$, of the mF-(n,0)-SWCNTs with that for the isolated (n,0)-SWCNTs for comparison with their oscillator strengths (f). Where m indicated the number of functional groups covalently bound to the (n,0)-SWCNTs and F symbolizes the benzenesulfonic acid used as functional group in this study.

$S_0 \rightarrow S_k$	(12,0)-SWCNT		4F-(12,0)-SWCNT		(8,0)-SWCNT		4F-(8,0)-SWCNT	
k	T_e(eV)	f	T_e(eV)	f	T_e(eV)	f	T_e(eV)	f
1	0.54	0.0076	0.12	0.0001	0.77	0.0164	0.09	0.0001
2	0.82		0.15	0.0001	1.46	0.0006	0.11	
3	0.82		0.16	0.0001	1.46	0.0006	0.14	
4	1.27		0.42		2.44		0.31	0.0003
5	1.51		0.55	0.0111	2.51	0.3588	0.36	0.0059
6	1.51		0.55	0.0113	2.51	0.3588	0.55	0.0013
7	1.71	0.1082	0.65	0.0355	2.53		0.62	
8	1.87	0.6641	0.65	0.0355	2.77		0.79	
9	1.87	0.6641	0.74	0.0098	2.77		0.93	0.0462
10	2.31		1.10	0.0054	2.78		1.86	0.0049
11	2.31		1.10	0.0053	2.78		1.91	
12	2.56		1.37	0.0072	2.98	0.1295	1.96	0.0022
13	2.76		1.37	0.0071			2.00	
14	2.76		1.49				2.09	0.0002
15	2.81		1.50				2.12	0.0010
16	2.82		1.52	0.0059			2.18	
17	2.82		1.55				2.21	0.0031
18	2.94		1.63	0.0138			2.24	0.0043
19	2.94		1.63	0.0137			2.27	0.0236
20	2.99		1.70	0.0014			2.29	0.0157

21	2.99		1.70	0.0013			2.30	
22	3.03		1.78				2.36	0.0018
23	3.03		1.89				2.38	
24	3.18	0.0301	2.05	0.2005			2.41	0.0015

$S_0 \rightarrow S_k$	(9,0)-SWCNT		2F-(9,0)-SWCNT		(7,0)-SWCNT		2F-(7,0)-SWCNT	
k	T_e(eV)	f	T_e(eV)	f	T_e(eV)	f	T_e(eV)	f
1	0.56		0.27		0.91		0.39	
2	0.80		0.30	0.0144	1.25		0.79	0.0573
3	0.80		0.57		1.25		1.07	0.0067
4	0.93	0.0391	0.83	0.0087	1.45	0.0647	1.11	
5	2.32		1.34		2.52			
6	2.32		1.46	0.0307	2.53			
7	2.50		1.47	0.1432	2.93			
8	2.50		1.57		2.93			
9	2.64		1.64		3.02			
10	2.64		1.84		3.02			
11	2.72		1.88		3.08			
12	2.72		1.88	0.0268	3.08			
13			2.10					
14			2.11	0.0020				
15			2.19					
16			2.24	0.0272				

5. STUDY OF POLYYNES ENCAPSULATED INTO SINGLE-WALLED CARBON NANOTUBE

One-dimensional carbon atomic wires displaying sp hybridization have an attractive electronic and vibrational structure which severely affects their optical and transport properties. These kinds of structure have received researchers' interest because of their purely *sp*-hybridized carbon structure that is expected to display a completely different behavior than the more common sp^2 and sp^3 carbon structures. Polyyne molecules are linear carbon chains having alternating single and triple bonds, and ended by end atoms or groups. A. Milani et al. [51] have investigated the charge transfer in carbon atomic wires (polyynes) terminated by phenyl rings and its effects on the structure of the system using normal Raman and surface-enhanced Raman spectroscopy (SERS) techniques as well density functional theory (DFT) calculations forthe

Raman modes. They reported that the occurrence of a charge transfer between polyynes and metal nanoparticles (both in liquids and supported on surfaces) is evidenced by Raman and SERS as a moderating of the vibrational stretching modes. They suggested that carbon wires alter their structure toward a more equalized geometry (i.e., all double bonds) as a consequence of the charge transfer. They also pointed out that these observations open potential perspectives for developing carbon-based atomic devices with tunable electronic properties. Therefore, it is necessary to carry out more experimental and theoretical investigation to get insight of them.

Even though the molecules like polyyne are very unstable at normal temperature and atmosphere conditions.[52, 53], it has beenreported that they are astoundingly stable inside single wall carbon nanotubes (SWCNT) even at high temperature (300 ^{o}C) [54, 55]. The Raman spectrum of the polyyne molecules exhibited two intense Raman shifts appear around 2000 -2200 cm^{-1}, which are labeled as α-bands and β-bands. The band positions of these two bands decrease in frequency with the increase in polyyne size. With the increasing chain lengths, while the frequency of the α-band almost linearly decreases, the position of β-bands is oscillating, and the difference between β-bands and α-bands in frequency shifts are dissimilar in polyyne molecules with different size.

Furthermore, L. M. Malard et al. [56] studied resonance Raman study of two polyyne molecules ($C_{10}H_2$ and $C_{12}H_2$) encapsulated inside the SWCNT using various different laser lines including the whole visible range. They indicated that the main Raman features associated with stretching modes of the linear chains in both samples ($C_{10}H_2$ @SWCNT and $C_{12}H_2$ @SWCNT) are strongly enhanced around 2.1 eV, while the optical absorption observed when these molecules are dispersed in isotropic medium [57] or in the gas phase[58] occurs above 4.5 eV. They concluded that dipole-forbidden (dark) transitions of the polyynes that become active as a result of a symmetry breaking when the molecules are encapsulated inside the SWCNT.

In this section, we will discuss the calculated results for the polyyne ($C_{10}H_2$) molecules encapsulated within (6,0)-SWCNT. Figure 18 and Table 6 provide the calculated electron density and energy levels of the molecular orbitals (MOs), HOMOs and LUMOs, of $C_{10}H_2$@(6,0)-SWCNT, respectively. The geometry optimization with/without symmetry restriction found the point group is respectively D_{6H} and D_{2H} symmetries. The structure with D_{2H} has the lowest energy as much as

0.19 eV than the structure with D_{6H} and both structure has the $^{1}A_{1G}$ electronic symmetry for the $C_{10}H_2$@(6,0)-SWCNT system.

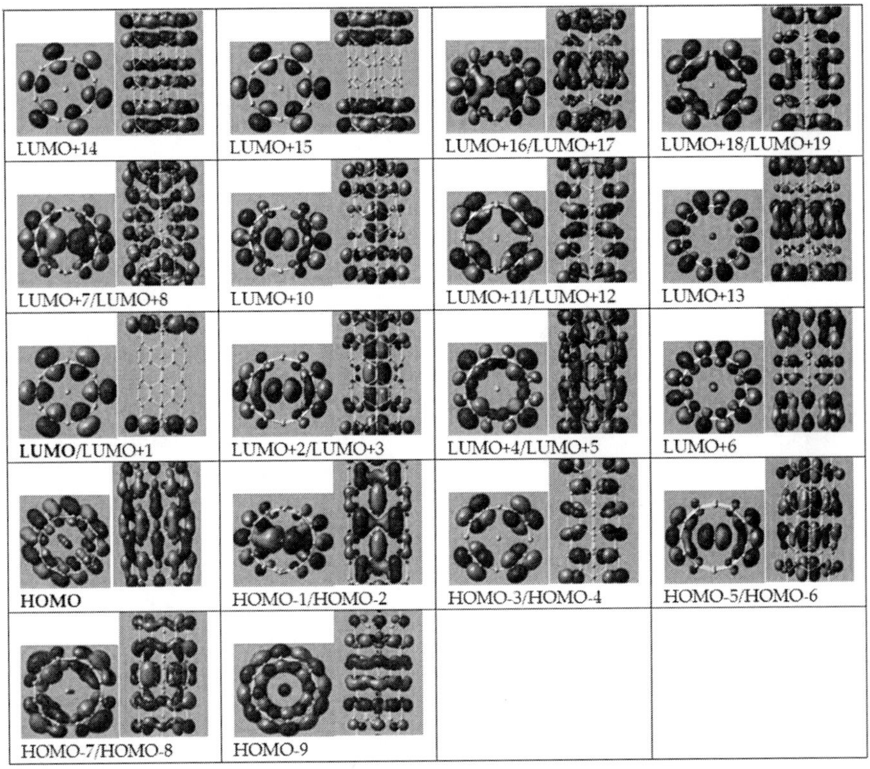

Figure 18. Calculated electron density of the molecular orbitals (MOs), HOMOs and LUMOs, of $C_{10}H_2$@(6,0)-SWCNT

For the isolated $C_{10}H_2$ (polyyne), predicted electronic symmetry is Σ11GΣ11G and has the $D_{\infty H}$ point group. As seen in Figure 18, the plotted electron density showed that while three of first five highest occupied molecular orbitals (HOMO/HOMO-3/HOMO-4 with the A_g, B_{1u} and A_{1u} symmetries, respectively) only belong to the (6,0)-SWCNT, the HOMO-1 and HOMO-2 with the B_{2u} and B_{3u} symmetry belong not only to both of the $C_{10}H_2$@(6,0)-SWCNT and but also there is a significant bonding interaction between the polyyne molecule ($C_{10}H_2$) and (6,0)-SWCNT in the ground state. As seen in Table 6, the lowest unoccupied molecular orbital, LUMO (B_{3u})/LUMO + 1(B_{2g})/LUMO + 4 (A_g) LUMO +

5(B_{1g}) and LUMO + 6(B_{1u}) and lies about 0.43/0.43/0.89/0.89 and 1.66 eV above the HOMO (A_g) belong to the SWCNT only and the LUMO + 1(B_{3u})/LUMO + 2 (B_{2u}) belongs to the polyyne molecule and the SWCNT. However, the LUMO + 7(B_{3g})/LUMO + 8(B_{2g}) and LUMO + 9(B_{3g}) belong not only to both the $C_{10}H_2$@(6,0)-SWCNT (lies 1.99 /1.99 and 2.39 eV above the HOMO (A_g)), but there is a significant sigma bonding interaction in the excited states as seen in Figure 16.

The bonding interactions between $C_{10}H_2$ and (6,0)-SWCNT in the ground state leading to the increase the triple bond lengths and decrease the double C-C bond lengths within the polyyne molecule ($C_{10}H_2$) when encapsulated inside the (6,0)-SWCNT relative to its corresponding bond distance of the isolated single polyyne chain molecules ($C_{10}H_2$).

For instance, C-C bond distances in the encapsulated $C_{10}H_2$ molecules: 1.25974, 1.26769, 1.33655, 1.24771, 1.35216, 1.24771, 1.33655, 1.26769, 1.25974 Å and corresponding C-C bond distances in the isolated one: 1.22161, 1.35656, 1.23246, 1.34527, 1.23515, 1.34527, 1.23246, 1.35656, 1.22161 Å. These σ-bonding interactions between $C_{10}H_2$ and (6,0)-SWCNTs in the ground and excited states may be aspirant for the charge transfer between the molecule and the SWCNT, which was observed between the polyyne and nanoparticles by the SERS as mentioned above.

The calculated vertical dipole allowed electronic transitions ($S_0 \rightarrow S_n$) of the $C_{10}H_2$@(6,0)-SWCNTs up to 0.52 eV are given in Table 7. Because of the technical difficulty, it was unable to calculate the higher electronic transitions that can provide more detailed information about internal conversion (IC) and inter system crossing (ISC). The lowest dipole allowed vertical electronic transitions $S_0(A_{1g}) \rightarrow S_7(B_{3u})$ as results of the HOMO-3→LUMO + 1 and HOMO→LUMO + 2 transitions and $S_0(A_{1g}) \rightarrow S_7(B_{2u})$ transition as a result of the HOMO-4->LUMO + 1 and HOMO→LUMO + 3 transitions,the second lowest dipole allowed vertical electronic transitions $S_0(A_{1g}) \rightarrow S_{11}(B_{3u})$ due to the HOMO-3→LUMO + 1 and HOMO→LUMO + 2 transitions, and $S_0(A_{1g}) \rightarrow S_{12}(B_{2u})$ transition because of the HOMO-4→LUMO + 1 and HOMO→LUMO + 3 transitions clearly indicate that the existence of charge transfer from the SWCNT to the polyyne molecules when examine the electron density of the HOMO and LUMOs involved in these transitions. When we examine the calculated vertical electronic transitions together with the calculated energy levels of molecular orbitals (MOs) of the encapsulated polyyne molecule inside the SWCNT, the IC and ISC can be expected.

Table 6. Calculated energy levels ΔE(eV) of the molecular orbitals (MOs) for the $C_{10}H_2$@(6,0)-SWCNTs and $C_{10}H_2$ relative to their the highest molecular orbital (HOMO)

$C_{10}H_2$@(6,0)-SWCNTs			$C_{10}H_2$	
MOs	Sym.	ΔE(eV)	Sym.	ΔE(eV)
LUMO + 14	B_{3g}	3.47	SG_u	13.87
LUMO + 13	A_g	2.86	PI_g	13.24
LUMO + 12	B_{1u}	2.72	PI_g	13.24
LUMO + 11	A_u	2.72	SG_g	12.57
LUMO + 10	B_{2g}	2.39	SG_u	11.56
LUMO + 9	B_{3g}	2.39	PI_u	11.07
LUMO + 8	B_{2g}	1.99	PI_u	11.07
LUMO + 7	B_{3g}	1.99	SG_g	9.16
LUMO + 6	B_{1u}	1.66	SG_u	9.15
LUMO + 5	B_{1g}	0.89	PI_g	8.55
LUMO + 4	A_g	0.89	PI_g	8.55
LUMO + 3	B_{2u}	0.87	PI_u	6.11
LUMO + 2	B_{3u}	0.87	PI_u	6.11
LUMO + 1	B_{2g}	0.43	PI_g	3.90
LUMO	B_{3u}	0.43	PI_g	3.90
HOMO	A_g	-0.00	PI_u	0.00
HOMO-1	B_{2u}	-0.19	PI_u	0.00
HOMO-2	B_{3u}	-0.19	PI_g	-1.53
HOMO-3	B_{1u}	-0.23	PI_g	-1.53
HOMO-4	A_u	-0.23	PI_u	-2.86
HOMO-5	B_{3g}	-1.55	PI_u	-2.86
HOMO-6	B_{2g}	-1.55	PI_g	-3.87
HOMO-7	A_g	-1.91		

Based on these calculations, the molecule encapsulated inside the nanotubes (NTs) can be used as energy conversion systems as a consequence of charge transfer between them. This illustration also can reflect on the intensity of the Raman bands at the resonance excitation energy where the charge transfer takes place between the molecules or particle and the nanotubes.

Table 7. The calculated vertical electronic transitions (Te; in eV) of $C_{10}H_2$@(6,0)-SWCNTs; S_0(A1g) → Sn. where the f and CI stand for the oscillator strength and the configurationally interaction coefficients, respectively. The Letters H and L stands for HOMO and LUMO, respectively.

		$S_0(A_{1G}) \to S_n$						$S_0 \to S_n$		
S_n	Sym.	H→L	CI	T_e(eV)	f	S_n	Sym.	H→L	CI	T_e(eV)
S_1	B_{2g}	H-"/>L+1	-0.84	0.06		S_{13}	B_{1g}	H-2-"/>L+3	-0.47	0.36
S_2	B_{3u}	H-"/>L	-0.83	0.06				H-1-"/>L+2	0.48	
S_3	B_{1g}	H-1-"/>L	0.61	0.08		S_{14}	A_g	H-2-"/>L+2	-0.49	0.37
S_4	A_g	H-2-"/>L	0.61	0.08				H-1-"/>L+3	0.49	
S_5	A_u	H-1-"/>L+1	0.62	0.09				H-"/>L+4	-0.15	
S_6	B_{1u}	H-2-"/>L+1	0.62	0.09		S_{15}	B_{1g}	H-2-"/>L+3	0.49	0.37
S_7	B_{3u}	H-3-"/>L+1	0.37	0.23	0.0001			H-1-"/>L+2	0.48	
		H-"/>L+2	0.56					H-"/>L+5	-0.15	
S_8	B_{2u}	H-4-"/>L+1	0.37	0.23	0.0001	S_{16}	A_g	H-2-"/>L+2	-0.13	0.41
		H-"/>L+3	0.57					H-1-"/>L+3	0.13	
S_9	B_{2g}	H-4-"/>L+3	-0.14	0.25				H-"/>L+4	0.60	
		H-3-"/>L	0.53			S_{17}	B_{1g}	H-2-"/>L+3	0.12	0.41
		H-3-"/>L+2	0.14					H-1-"/>L+2	0.12	
S_{10}	B_{3g}	H-4-"/>L	0.53	0.25				H-"/>L+5	0.60	
		H-4-"/>L+2	-0.14			S_{18}	B_{2g}	H-4-"/>L+2	-0.49	0.50
		H-3-"/>L+3	-0.14					H-3-"/>L+3	0.49	
S_{11}	B_{3u}	H-3-"/>L+1	0.40	0.30	0.0012	S_{19}	B_{2u}	H-2-"/>L+5	-0.51	0.51
		H-"/>L+2	-0.39					H-1-"/>L+4	0.52	
S_{12}	B_{2u}	H-4-"/>L+1	0.40	0.30	0.0012	S_{20}	A_u	H-4-"/>L+4	-0.44	0.52
		H-"/>L+3	-0.39					H-3-"/>L+5	0.44	

REFERENCES

1. S. Iijima, 1991 Nature. 354 56

2. Nepal. Dhriti, Sohn. Jung-Inn, K. Aicher Wilhelm, Lee. Seonghoon, E. Geckeler Kurt, 2005 Biomacromolecules 6 6 2919

3. S. S. Karajanagi, H. Yang, P. Asuri, E. Sellitto, J. S. Dordick, R. S. Kane, 2006 Langmuir 22 1392

4. Finkelstein. Hod, M. Asbeck Peter, Esener Sadik, 2003 3rd IEEE Conference on Nanotechnology (IEEE-NANO) 1 441

5. C. Zhou, J. Kong, E. Yenilmez, H. Dai, 2000 Science 290 1552

6. J. A. Misewich, R. Martel, P. Avouris, J. C. Tsang, S. Heinze, J. Tersoff, 2003 Science 300 783

7. A. Nitzan, R. A. Ratner, 2003 Science 300 1384

8. D. S. Wen, Y. L. Ding, 2004 International Journal of Heat and Mass Transfer 47 5181

9. H. Masuda, A. Ebata, K. Teramae, N. Hishiunma, 1993 Netsu Bussei (Japan) 4 227

10. S. Jones Carissa, Lu Xuejun, Renn Mike, Stroder Mike, Shih. Wu-Sheng, 2009 Microelectronic Engineering DOI: j.mee.2009.05.034

11. S. M. Jung, H. Y. Jung, J. S. Suh, 2009 Sensors and ActuatorsB. 139 425

12. T. Rueckes, K. Kim, E. Joselevich, G. Y. Tseng, C. L. Cheung, C. M. Lieber, 2000 Science 289 94

13. J. O'Connell Michael, M. Bachilo Sergei, B. Huffman Chad, C. Moore Valerie, S. Strano Michael, H. Haroz Erik, Kristy. L. Rialon, Peter. J. Boul, William. H. Noon, Kittrell Carter, Jianpeng. Ma, H. Hauge Robert, R. Bruce Weisman, E. Richard Smalley, 2002 Science 297 593

14. S. M. Bachilo, M. S. Strano, C. Kittrell, R. H. Hauge, R. E. Smalley, R. B. Weisman, 2002 Science 298 2361

15. A. Hartschuh, H. N. Pedrosa, L. Novotny, T. D. Krauss, 2003 Science 301 1354

16. J. Maultzsch, R. Pomraenke, S. Reich, E. Chang, D. Prezzi, A. Ruini, E. Molinari, M. S. Strano, C. Thomsen, C. Lienau, 2006 Phys. Stat. Sol. B 243 13 3204

17. E. Chang, G. Bussi, A. Ruini, E. Molinari, 2004 Phys. Rev. Lett. 92 196401

18. C. D. Spataru, S. Ismail-Beigi, L. X. Benedict, S. G. Louie, 2004 Phys. Rev. Lett. 92 077402

19. V. Perebeinos, J. Tersoff, P. Avouris, 2004 Phys. Rev. Lett. 92 257402

20. Park Won-Il, Hun-Sik Kim, Soon-Min Kwon, Young-Ho Hong, Hyoung-Joon Jin, 2009 Carbohydrate Polymers 77 457

21. Lingjie Meng, Chuanlong Fu, Qinghua Lu, 2009 Natural Science 19 801

22. C. T. Hsieh, Y. T. Lin, 2006 Microporous Mesoporous Mater. 93 232

23. J. J. Davis, K. S. Coleman, B. R. Azamian, C. B. Bagshaw, M. L. H. Green, 2003Chem Eur J. 9 16 3732

24. V. Z. Poenitzsch, D. C. Winters, H. Xie, G. R. Dieckmann, A. B. Dalton, I. H. Musselman, 2007 J Am ChemSoc, 129 47 14724

25. Carbon nanotube science: 2009 Synthesis, Properties and Applications, by P.J.F. Harris (Cambridge University Press, Cambridge).

26. Liu Zunfeng, et al. 2010 Journal of Nanoscience and Nanotechnology 10 5570

27. Robel Istvan, A Bruce, Prashant V. Kamat, 2005 Advanced Materials 17 20 2458

28. M. Olek, T. Busgen, M. Hilgendorff, M. Giersig, 2006 Journal of Physical Chemistry B 110 26 12901

29. Yanagi Kazuhiro, Hiroyuki Matsui Konstantin Iakoubovskii, Matsuzaki Hiroyuki, Okamoto Hiroshi, Miyata Yasumitsu, Maniwa Yutaka, Kazaoui Said, Minami Nobutsugu, Kataura Hiromichi, 2007 J. Am. Chem. Soc. 129 16 4992

30. Saito Yuika, Yanagi Kazuhiro, Hayazawa Norihiko, Ishitobi Hidekazu, Ono Atsushi, Kataura Hiromichi, Kawata Satoshi, 2006 Jap. J. Appl. Phys. 45 12 9286

31. L. Alvarez, Y. Almadori, R. Arenal, R. Babaa, T. Michel, R. Le Parc, J. -L Bantignies, B. Jousselme, S. Palacin, P. Hermet, J. L. Sauvajol, 2011 J. Phys. Chem. C, 115 24 11898

32. C. Srinivasan, 2008 Current Science 94 300

33. T. A. Hilder, J. M. Hill, 2008 Current Applied Physics8 258

34. J. Clendenin, J. Kim, S. Tung, 2007 Proc of 2007 2nd IEEE conference on Nanotechnology 1028

35. A. Bianco, K. Kostarelos, M. Prato, 2005 Current Opinion in Biotechnology 9 674

36. G. Nasreen, R. J. Chopra, K. Luyken, Vincent. H. Cherrey, Marvin. L. Crespi, Steven. G. Louie Cohen, A. Zettl, 1995Science. 269 966

37. D. Golberg, Y. Bando, W. Han, K. Kurashima, T. Sato, 1999 Chem. Phys. Lett. 308 337

38. C. C. Tang, Y. Bando, T. Sato, K. Kurashima, 2002 Chem. Commun. 121 290

39. M. S. Dresselhaus, G. Dresselhaus, P. C. Eklund, 1996 Science of Fullerenes and Carbon NanotubesAcademic Press, San Diego, CA.

40. A. Rubio, J. Corkill, M. L. Cohen, 1994 Phys. Rev. B. 49 5081

41. D. Zhang, R. Q. Zhang, 2003Chem. Phys. Lett. 371 426

42. M. Mirzaei, N. L. Hadipour, 2008 Physica E, 40 800

43. C. H. Park, C. D. Spataru, S. G. Louie, 2006 Physical Review Letters 96 126105

44. L. Wirtz, A. Marini, A. Rubio, 2006 Physical Review Letters 96 126104

45. Zhi Chunyi, Bando Yoshio, Tang Chengchun, Golberg Dmitri, 2010 Materials Science and Engineering R 70 92

46. N. W. S. Kam, M. O'Connell, J. A. Wisdom, H. Dai, 2005 PNAS 102 11600

47. Gaussian, Inc. Carnegie Office Park-Bulding 6, Pittsburgh, PA106, USA.

48. (a) C. Guo, M. Aydin, H. R. Zhu, D. L. Akins, J. Phys. Chem. B 106 (2002) 5447; (b) H, Guo, X. Zhang, M. Aydin, W. Xu, H. R. Zhu, D. L Akins, Journal of Molecular Structure 689 (2004) 153; (c) M. Aydin, F. Jean-Mary, N. Stevens, D. L. Akins, J. Phys. Chem. B 108(2004) 9695; (d) W. Xu, M. Aydin, S. Zakia, D. L. Akins, J. Phys. Chem. B 108 (2004) 5588; (e) M. Aydin, J. R. Lombardi, J. Phys. Chem. A. 113 (2009) 2809; (f) M. Aydin, D.L. Akins, Vibrational Spectroscopy 53 (2010) 163; (g) M. Aydin, Vib. Spectrosc. 66 (2013) 30, (h) Metin Aydin and Daniel Akins (2011), Geometric and Spectroscopic Properties of Carbon Nanotubes and Boron Nitride Nanotubes, Electronic Properties of Carbon Nanotubes, Prof. Jose Mauricio Marulanda (Ed.), ISBN: 978-953-307-499-3, InTech, Available from: http://www.intechopen.com/books/electronic-properties-of-carbon-nanotubes/geometric-and-spectroscopic-properties-of-carbon-nanotubes-and-boron-nitride-nanotubes; (i) Metin AYDIN Photofragmentation Spectroscopy, Publisher: VDM Verlag, 2009; ISBN: 3639152891.

49. R. Pfeiffer, F. Simon, H. Kuzmany, V. N. Popov, 2005 Phys. Rev. B72 161404(R)

50. Eun-Ju Lee, Jin-San Yoon, Mal-Nam Kim, Eun-Soo Park, 1970 Preparation and Applicability of Vinyl Alcohol Group Containing Polymer/MWNT Nanocomposite Using a Simple Saponification Method Carbon Nanotubes- Polymer Nanocomposites, Siva Yellampalli (Ed.) 978-9-53307-498-6 InTech, Available from: http://www.intechopen.com/books/carbon-nanotubes-polymer-nanocomposites/preparation-and-applicability-of-vinyl-alcohol-group-containing-polymer-mwnt-nanocomposite-using-a-s

51. A. Milani, A. Lucotti, V. Russo, M. Tommasini, F. Cataldo, A. Li Bassi, C. S. Casari, 2011 J. Phys. Chem. C 115 12836

52. L. Ravagnan, F. Siviero, C. Lenardi, P. Piseri, E. Barborini, P. Milani, C. S. Casari, A. Li Bassi, C. E. Bottani, 2002 Phys. Rev. Lett. 89285506

53. C. S. Casari, A. Li Bassi, L. Ravagnan, F. Siviero, C. Lenardi, P. Piseri, G. Bongiorno, C. E. Bottani, P. Milani, 2004 Phys. Rev. B 69075422

54. Nishide Daisuke, Dohi Hirofumi, Wakabayashi Tomonari, Nishibori Eiji, Aoyagi Shinobu, Ishida Masashi, Kikuchi Satoshi, Kitaura Ryo, Sugai Toshiki, Sakata Makoto, Shinohara Hisanori, 2006 Chem. Phys. Letter 428 356

55. D. Nishide, H. Dohi, T. Wakabayashi, E. Nishibori, S. Aoyagi, M. Ishida, S. Kikuchi, R. Kitaura, T. Sugai, M. Sakata, H. Shinohara, 2006 Chem. Phys. Lett. 428 356

56. L. M. Malard, D. Nishide, L. G. Dias, B. Capaz Rodrigo, A. P. Gomes, A. Jorio, C. A. Achete, R. Saito, Y. Achiba, H. Shinohara, M. A. Pimenta, 2007 Phys. Rev. B 76 233412

57. H Tabata, M Fujii, S Hayashi, T Doi, T Wakabayashi, 2006 Carbon 44 3168 T Wakabayashi, H Tabata, T Doi, H Nagayama, K Okuda, R Umeda, I Hisaki, M Sonoda, Y Tobe, T Minematsu, K Hashimoto, S Hayashi, 2007 Chem. Phys. Lett. 433 296

58. T Pino, H Ding, F Güthe, J. P. Maier, 2001 J. Chem. Phys. 114 2208

59. Cumings, A. Zettl, Chem. Phys. Lett. 316 (2000) 211.

60. N. R. Arutyunyan, E. D. Obraztsova, M. Silly, P. Jaffrennou, B. A. Tretout, A. Loiseau, A. L. Chuvilin, Phys. Stat. Sol. (b) 243(13) (2006) 3316.

61. S. Y. Bae, H. W. Seo, J. Park, Y. S. Choi, J. C. Park, S. Y. Lee, Chem. Phys. Lett. 374 (2003) 534.

62. W.-Q. Han, H.-G. Yu, C. Zhi, J. Wang, Z. Liu,T. Sekiguchi, Y. Bando, Nano Lett. 8(2) (2008) 491.

63. E. A. Obraztsova, D. V. Shtansky, A. N. Sheveyko, M. Yamaguchi, A. M. Kovalskii, D. Golberg, Scripta Mater. 67 (2012) 507.

64. C. H. Lee, J. Wang, V. K. Kayatsha, J. Y. Huang, Y. K. Yap, Nanotechnology 19 (2008) 455605.

6

EFFICIENT NEAR-INFRARED UP-CONVERSION PHOTOLUMINESCENCE IN CARBON NANOTUBES

Naoto Akizuki[1], Shun Aota[1], Shinichiro Mouri[1], Kazunari Matsuda[1] & Yuhei Miyauchi[1,2,3]

[1]Institute of Advanced Energy, Kyoto University, Uji, Kyoto 611-0011, Japan.

[2]PRESTO, Japan Science and Technology Agency, 4-1-8 Honcho Kawaguchi,

Saitama 332-0012, Japan.

[3]Graduate School of Science, Nagoya University, Chikusa, Nagoya 464-8602, Japan.

Photoluminescence phenomena normally obey Stokes' law of luminescence according to which the emitted photon energy is typically lower than its excitation counterparts. Here we show that carbon nanotubes break this rule under one-photon excitation conditions. We found that the carbon nanotubes exhibit efficient near-infrared photoluminescence upon photoexcitation even at an energy lying >100–200 meV below that of the emission at room temperature. This apparently anomalous phenomenon is attributed to efficient one-phonon-assisted up-conversion processes resulting from unique excited-state dynamics emerging in an individual carbon nanotube with accidentally or intentionally embedded localized states. These findings may open new doors for energy harvesting, optoelectronics and deep-tissue photoluminescence imaging in the near-infrared optical range.

Photoluminescence phenomena normally obey Stokes' law of luminescence[1], which states that the emission light wavelength (energy) is typically longer (lower) than its excitation counterpart. Up-conversion (or anti-Stokes) photoluminescence[2] violates the Stokes' law of luminescence owing to additional energy gain induced by multiple photon[2, 3, 4, 5] or thermal (phonon) energy[6, 7]absorption. This unique photon up-conversion phenomenon only occurs efficiently in certain materials such as rare-earth-doped materials[2, 6, 8] and some organic dyes[3, 4] by virtue of their specific electronic structures. Its potential usefulness in various applications such as photovoltaic energy conversion[3, 9], photoluminescence bio-imaging[8, 10, 11, 12, 13], displays[14], lasers[15] and optical refrigeration of solids[6, 16] has attracted growing interest in recent years. However, especially in the near-infrared optical range, limitations in materials and combinations of excitation and emission photon energies available for efficient photon up-conversion have reduced its compatibility with a wide range of applications.

Single-walled carbon nanotubes (hereafter referred to as carbon nanotubes) consist of graphene sheets rolled up into quasi-one-dimensional tubular structures with diameters of the order of 1 nm (ref. 17). Depending on the rolling-up direction and their diameter defined by two indices (n, m)[18], nanotubes act as direct gap semiconductors with band gaps of the order of 1 eV and emit photoluminescence in the near-infrared optical range[19] because of recombination of excitons (electron and hole pairs bound via attractive Coulomb interactions)[20, 21, 22]. Since the first observation of efficient Stokes photoluminescence from individual carbon nanotubes[19, 23], numerous distinctive luminescence properties of nanotubes and their applications have been uncovered over the past decade[24, 25, 26, 27]. However, this report reveals an additional remarkable characteristic of carbon nanotubes as photoluminescent materials.

Here, we demonstrate that carbon nanotubes[17] are new up-conversion phosphors showing efficient near-infrared to near-infrared up-conversion photoluminescence at room temperature with an energy gain of up to a few hundreds of milli-electronvolts, corresponding to light wavelength shortening of a few hundreds of nanometres. This photon up-conversion does not originate from common coherent two-photon absorption or anti-Stokes Raman processes. The observed phenomena are attributed to efficient excitonic up-conversion processes gaining energy by absorbing high-energy (~100–200 meV) optical phonons characteristic in carbon nanotubes, although the phonon energy much exceeds the ambient thermal energy (only ~26 meV) and corresponding phonon number available for the up-conversion is considerably small. This apparently anomalous process stems from unique excited-state dynamics emerging in an individual carbon nanotube with accidentally or intentionally embedded localized states. These findings may lead to new opportunities for near-infrared and thermal energy harvesting, optoelectronics and deep-tissue up-conversion photoluminescence imaging using conventional silicon-based image sensors and low-cost near-infrared light sources.

1. RESULTS

1.1 Optical imaging of the up-conversion light emission

Figure 1a shows an optical image of photoemission from (6,5)-rich carbon nanotubes dispersed in D_2O (Methods section) in a standard quartz cell. The incident photon wavelength (energy) was set to ca. 1,100 nm (1.13 eV) and the photoemission image ranging between 950 and 1,000 nm (1.24–1.31 eV) was selectively captured using optical filters. The power density of the collimated incident light equalled 200 mW cm^{-2}, which was too weak to induce coherent two-photon excitation photoluminescence[21, 22]. Although the incident photon energy (wavelength) was considerably lower (longer) than the detected photon energy (wavelength), photoemission was clearly observed along the incident light path. We also confirmed that this light emission is from individual carbon nanotubes by directly observing microscopic light emission image of an individual nanotube under the up-conversion excitation condition as shown in Fig. 1b. This situation is schematized in Fig. 1c.

1.2 Up-conversion photoemission spectra

Figure 2a compares photoemission spectra of nanotubes excited with 1.13 (1,100 nm; red curve) and 2.18 eV (568 nm; black curve) photons, respectively. The grey curve shows control data obtained for a nanotube-free surfactant solution excited at 1.13 eV. The emission peak detected at ca. 1.26 eV upon excitation at 2.18 eV (resonant with the second sub-band exciton energy $E_{22}$of (6,5) nanotubes) corresponds to the photoluminescence induced by exciton recombination in the first sub-band (E_{11}) of (6,5) nanotubes[23]. The photoemission spectrum obtained at an excitation energy of 1.13 eV exhibited a peak at ca. 1.26 eV. The similar peak energies for excitation energies of 1.13 and 2.18 eV indicate that the photoemission mainly originates from (6,5) nanotubes (Supplementary Note 1 and Supplementary Fig. 1).

1.3 Excitation power dependence

The excitation power density dependence was examined to clarify the origin of the observed up-conversion photoemission. The spectral shape hardly changed with excitation power density variations (Fig. 2b). Figure 2c shows the up-conversion photoemission intensity as a function of excitation power density, revealing a weak sublinear dependence between these parameters. These observations exclude any possible photoemission induced by multiple-photon absorption (coherent or incoherent) or Auger recombination-mediated up-conversion[28] as origins for the observed phenomena because these mechanisms usually display a superlinear (quadratic or more) excitation power dependence.

1.4 Excitation photon energy dependence

Next, the excitation photon energy dependence of the up-conversion photoemission spectra was evaluated (Fig. 2d). The up-conversion intensity strongly depended on the excitation photon energy (wavelength) but spectral shape and peak positions remain nearly intact upon excitation photon energy variations. This observation rules out the possibility that the up-conversion photoemission results from anti-Stokes Raman scattering because Raman processes induce clear peak shifts according to excitation energy changes.

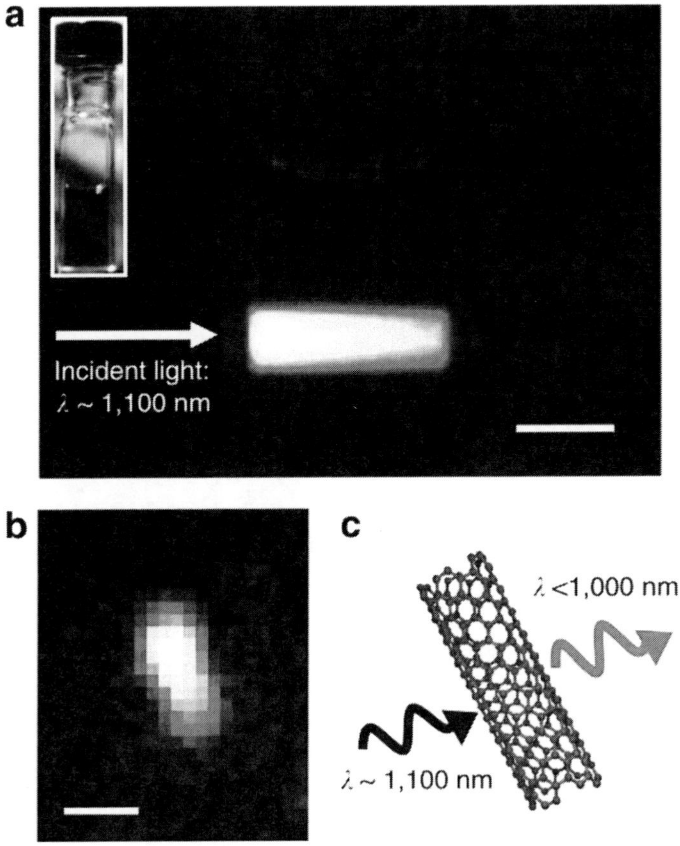

Figure 1. Up-conversion light emission of carbon nanotubes. (**a**) Optical image showing light emission from (6,5)-rich carbon nanotubes dispersed in D_2O using surfactant (sodium dodecyl benzene sulfonate, Methods section) excited by incident light of wavelength ~1,100 nm (selected wavelength range of 1,075–1,125 nm, power density=200 mW cm^{-2}, collimated). The photoemission image was selectively detected in the 950–1,000 nm wavelength range using optical filters. The micelle-suspended nanotube dispersion was placed in a quartz cell (inset). Scale bar, 5 mm. (**b**) Microscopic image of the up-conversion light emission from an individual nanotube taken under the same excitation wavelength (~1,100 nm) used in the experiment shown in **a**. Approximate excitation power density was 4 kW cm^{-2} (Methods section). Scale bar, 1 μm. (**c**) Optical processes observed in the experiment shown in **a** and **b**. A shorter wavelength (higher energy) photoemission (<1,000 nm) results from excitation at longer wavelength (lower energy) incident light (~1,100 nm).

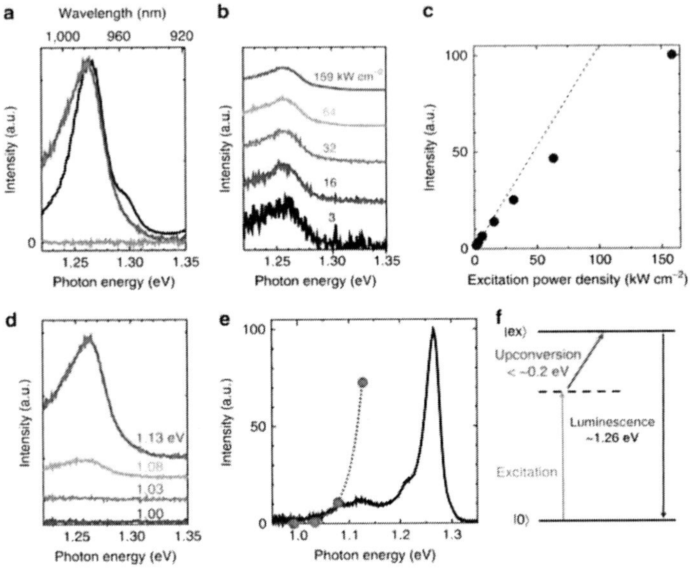

Figure 2. Excitation power and energy dependences of the up-conversion photoemission. (**a**) Photoemission spectra of carbon nanotubes excited at incident photon energies of 1.13 (1,100 nm, red curve) and 2.18 eV (568 nm, black curve). The gray curve is a spectrum taken for a carbon nanotube-free surfactant solution excited at 1.13 eV for comparison. The intensity was normalized so that the noise level appears to be comparable to the spectrum of carbon nanotubes excited at 1.13 eV (red curve). The excitation band width was 10 nm. (**b**) Excitation power density dependence of photoemission spectra excited at 1.13 eV. The spectra for excitation densities from 3 to 159 kW $cm^{-2}$are shown. Each spectrum is normalized by corresponding excitation power density. (**c**) Photoemission intensity (from data shown in **b**) plotted as a function of excitation power density. The dotted line corresponds to linear function for comparison. (**d**) Excitation photon energy dependence of the photoemission spectra. In **b** and **d**, each spectrum was vertically shifted for comparison, and corresponding excitation power densities (for **d**) or excitation photon energies (for **d**) are indicated on each spectrum for clarity. (**e**) Up-conversion photoemission intensity (red circles, from spectra in **d**) as a function of excitation photon energy, plotted together with normal photoluminescence spectrum at an excitation energy of 2.18 eV (black curve) for comparison. The dotted red curve is a guide to the eye. (**f**) Energy diagram for the observed photon up-conversion processes. Solid lines represent the ground (|0>) and free exciton (|ex>) states in carbon nanotubes. The dashed line corresponds to an intermediate state in the up-conversion process.

The exclusion of these potential processes suggests that the observed up-converted photoemission is attributable to photoluminescence induced by one-photon absorption followed by an internal excitonic up-conversion process in a carbon nanotube. Therefore, the observed phenomena are referred to as up-conversion photoluminescence in the remainder of the manuscript. Let us now consider the mechanism that governs this efficient excitonic up-conversion. The excitation photon energy dependence of the up-conversion intensity may provide a clue to the mechanism. Figure 2e compares the observed up-conversion photoluminescence intensity (red circles) as a function of excitation photon energy with a normal (Stokes) photoluminescence spectrum (black curve) obtained at an excitation energy of 2.18 eV. The threshold excitation photon energy required to achieve an observable up-conversion photoluminescence reaching ca. 1.3 eV amounted to 1.03–1.08 eV. A diagram showing excitation and emission photon energies for the observed up-conversion processes is shown in Fig. 2f.

1.5 Temperature dependence

The threshold energy gain close to the maximum optical phonon energy (~0.2 eV) in carbon nanotubes[29] and increasing up-conversion efficiency for smaller energy gain imply that phonon-assisted processes may induce internal up-conversion in carbon nanotubes. Therefore, temperature-dependent studies on the up-conversion photoluminescence were conducted. Figure 3a compares up-conversion photoluminescence intensities at 297 and 281 K at an excitation energy of 1.13 eV. The up-conversion photoluminescence intensity exhibited high sensitivity to small temperature variations and increased with increasing temperature. Figure 3b shows the up-conversion (red open circles) and normal photoluminescence intensities (black open circles) as functions of temperature. Up-conversion and normal photoluminescence intensities displayed opposite trends with respect to temperature (Fig. 3b). The up-conversion photoluminescence intensity increased, whereas its normal counterpart slightly decreased when the temperature increased.

Figure 3c shows an Arrhenius plot of the up-conversion intensity (I_{UCPL}) divided by the normal photoluminescence intensity (I_{PL}) at each corresponding temperature. The solid line in Fig. 3crepresents the Arrhenius equation ($\propto \exp(-E_a/k_B T)$ where E_a, k_B and T are the activation energy, Boltzmann constant and temperature) for an E_a value of 120 meV. (See also Supplementary Fig. 2 for data obtained at an excitation energy of 1.08 eV). The experimental trend was well reproduced with an E_a value closely matching the typical energy gain (120–130 meV) at an excitation energy of 1.13 eV, strongly suggesting that the internal exciton up-conversion originates from phonon absorption processes in carbon nanotubes, rather than other possible processes such as triplet–triplet annihilation[3, 30]. Note that the activation energy $E_a$nearly equal to the energy gain indicates one-step up-conversion via absorption of an optical phonon with much higher energy (~100–200 meV) than k_BT (~26 meV), distinct from common thermal activation process in which the additional energy is mainly provided by absorption of multiple phonons with energy comparable to or lower than k_BT.

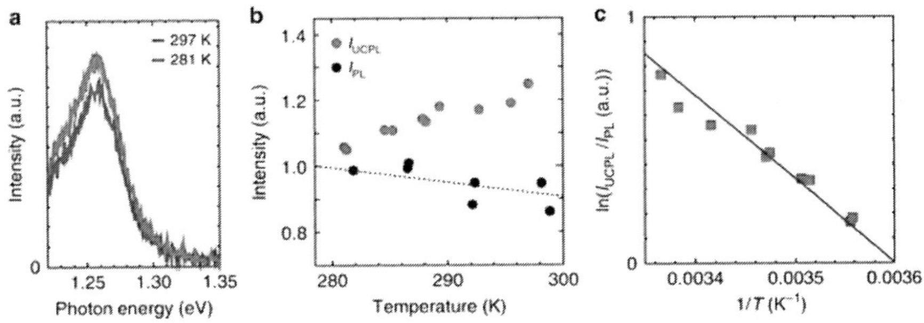

Figure 3. Temperature dependence of the up-conversion photoluminescence. (**a**) Up-conversion photoluminescence spectra at 297 (red curve) and 281 K (blue curve) for an excitation energy of 1.13 eV. (**b**) Up-conversion (red circles, excitation energy=1.13 eV, I_{UCPL}) and normal photoluminescence intensities (black circles, excitation energy=2.18 eV, I_{PL}). The temperature-dependent normal photoluminescence intensity is fitted with a linear function (dotted line), which was used to normalize the up-conversion photoluminescence intensity at each temperature in **c**. (**c**) Arrhenius plot of the up-conversion intensity (I_{UCPL}) normalized by the normal photoluminescence intensity (I_{PL}) at the corresponding temperature (red squares). I_{UCPL} was divided by I_{PL} to account for the temperature-dependent variation of I_{PL}. The black solid line represents the Arrhenius equation for an activation energy of 120 meV.

1.6 Intermediate states in up-conversion processes

Now that the up-conversion photoluminescence has been clearly attributed to phonon-assisted processes, the reason for the efficient phonon-assisted up-conversion with an energy gain reaching 5–10 k_BT needs to be elucidated. The weak sublinear dependence of the up-conversion photoluminescence intensity on the excitation power density (Fig. 2c) may provide insight into this anomalous phenomenon. The sublinear dependence implies the existence of an intermediate state with finite lifetime causing state-filling-induced saturation effect. Because it lies at lower energy than the E_{11} free exciton state, this state may be a mid-gap localized state[31, 32, 33, 34]unintentionally generated in carbon nanotubes during sample preparation because some types of these states[32, 33, 34, 35, 36] are optically active and can be photoexcited directly by near-infrared photons.

To test this hypothesis, localized states were intentionally embedded in carbon nanotubes by chemical treatments[32, 33, 34], and up-conversion photoluminescence changes were examined. Carbon nanotubes with oxygen- and sp^3-defect-derived optically active localized (zero-dimension (0D)-like) states were prepared according to procedures originally developed by Ghosh *et al.*[32]and Piao *et al.*[34] with some modifications in selections of chemicals and experimental parameters, respectively (Methods section). Figure 4a,f compare the normal photoluminescence spectra of as-dispersed (black curves) and sp^3-defect (red curve) or oxygen-doped (blue curve; hereafter referred to as 0D-doped) carbon nanotubes. A relatively broad and intense photoluminescence band emerged around 1.1 eV after chemical treatment, whereas the originalE_{11} photoluminescence intensities decreased only slightly for 0D-doped nanotubes. Figure 4b–e,g–j show up-conversion photoluminescence spectra of as-dispersed (black curves) and 0D-doped carbon nanotubes (coloured curves) excited at 1.13, 1.08, 1.03 and 1.00 eV, respectively. Note that the 0D-doped nanotubes displayed a considerably enhanced up-conversion photoluminescence intensity compared with their as-dispersed counterparts regardless of origins of the localized states.

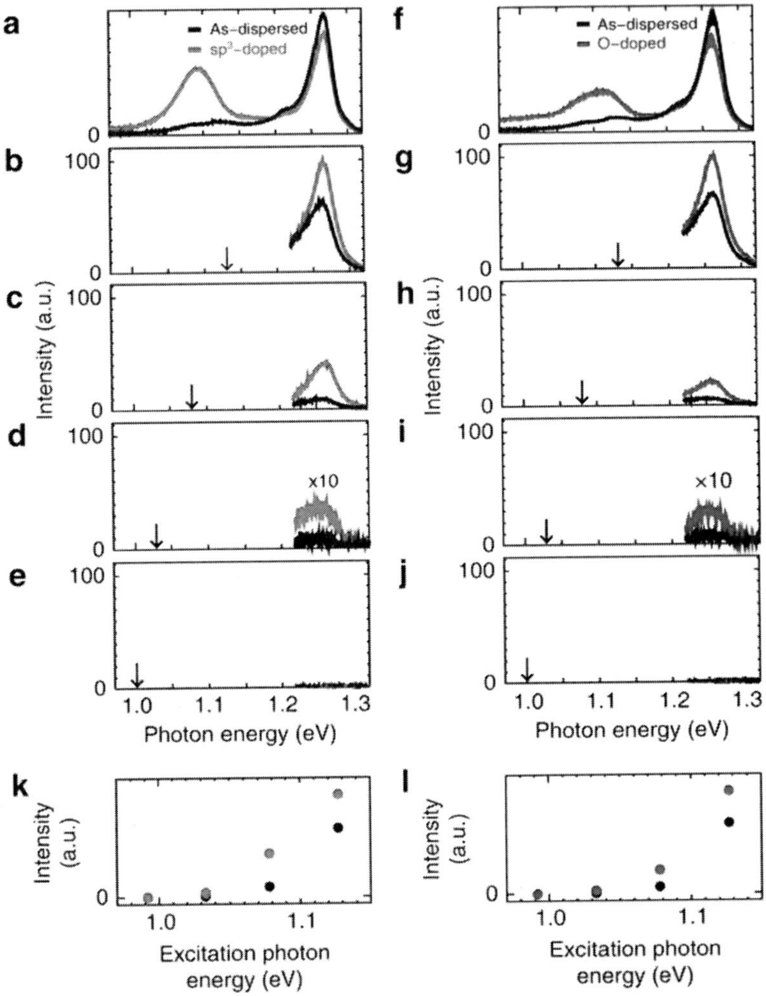

Figure 4. Up-conversion photoluminescence with intentionally doped states. (**a**,**f**) Photoluminescence spectra of as-dispersed (black curves) and 0D-doped carbon nanotubes (red and blue curves for (**a**) sp^3-and (**f**) oxygen-doped structures, respectively). **a** and **f** show normal photoluminescence spectra of each sample excited at 2.18 eV. (**b**–**e**, **g**–**j**) Up-conversion photoluminescence spectra of samples excited at (**b**,**g**) 1.13, (**c**,**h**) 1.08, (**d**,**i**) 1.03 and (**e**,**j**) 1.00 eV, respectively. Vertical arrows in each panel indicate the corresponding excitation photon energy. Spectra in **d**,**i** for an excitation energy of 1.03 eV were magnified 10 times. (**k**,**l**) Excitation energy-dependent up-conversion photoluminescence intensities of as-dispersed (black circles) and 0D-doped (red and blue circles for sp^3- and oxygen-doped structures, respectively) carbon nanotubes.

Figure 4k,l compare the excitation energy-dependent up-conversion photoluminescence intensities of as-dispersed and 0D-doped nanotubes. The up-conversion photoluminescence intensity increased dramatically at the excitation energy of 1.08 eV for 0D-doped nanotubes. Because this energy closely matched the photoluminescence peak energy of the localized excitons (Fig. 4a,f), this enhancement is attributable to exciton up-conversion from the additional localized states. The temperature (Supplementary Fig. 3) and the excitation power density dependences (Supplementary Note 2 and Supplementary Fig. 4) of the up-conversion photoluminescence were also assessed for the 0D-doped nanotubes. The results were similar to those obtained for as-dispersed nanotubes. This strongly suggests that the embedded localized states act as intermediate states in up-conversion processes. Moreover, the strong enhancement of the up-conversion photoluminescence achieved by simple chemical treatments is attractive for their applications as up-conversion phosphors.

2. DISCUSSION

We deduced the up-conversion quantum yield of the localized excitons to the free excitons using the difference between up-conversion photoluminescence intensities before and after intentional localized state doping (see Supplementary Discussion and Supplementary Fig. 9 for details). It was of the order of 10^{-1} for an energy gain of 120 meV even at room temperature. This highly efficient phonon-assisted exciton up-conversion showing an energy gain much exceeding the thermal energy appears anomalous but may be understood as a consequence of rapid exciton migration along the nanotube axis[37] as described below. The efficient exciton scattering by acoustic phonons[38] or random extrinsic surface potentials[39] that drive rapid diffusive exciton migration[37, 40, 41, 42, 43, 44] surpassing >100 nm along the nanotube axis spatially separates the up-converted excitons from their original localized state immediately after up-conversion and strongly reduces their probability of returning to this original state (see Supplementary Discussion and Supplementary Figs 10–12 for details). This unique exciton dynamics emerging in a carbon nanotube with embedded 0D-like localized states enables efficient phonon-assisted exciton up-conversion with an energy gain exceeding the thermal energy by one order of magnitude at room temperature. A potential enhancement of the exciton–phonon coupling resulting from exciton localization[34, 45, 46] may also contribute to the efficient phonon-assisted exciton up-conversion in carbon nanotubes.

Finally, let us discuss the merits of the up-conversion excitation wavelength of the nanotubes lying in a specific near-infrared optical range (λ=1,000–1,400 nm). This wavelength range is known as 'second biological transparency window' (NIR II), in which the light penetration depth into biological tissues is maximized, whereas photo-induced degradation and autofluorescence are minimized[25, 47]. Indeed, carbon nanotubes are expected to serve as promising phosphors for NIR II deep-tissue imaging, and biocompatible and luminescent nanotube dispersion has already been developed[25, 47]. A chief obstacle to this imaging approach has been the unsuitability of common silicon-based image sensors because of their vanishing sensitivity for NIR II photons. The up-conversion photoluminescence of the (6,5) carbon nanotubes with the inverted excitation (in the NIR II range) and emission (<1,000 nm, so-called NIR I range) wavelengths may solve this problem and facilitate deep-tissue bio-imaging by taking advantage of the large penetration depth of NIR II photons and common silicon-based image sensors (Supplementary Notes 3–5 andSupplementary Figs 5–8). Because up-conversion occurs as a one-photon process in carbon nanotubes, it does not require costly ultrafast high power lasers and high-density excitation conditions unlike conventional multi-photon excitation imaging. Furthermore, up-converted photocarriers in carbon nanotubes may be directly extracted as electronic current by configuring photovoltaic devices because of their quasi-one-dimensional structures. Therefore, these findings may also lead to new routes for optoelectronics and heat and photon harvesting below the optical gap in the near-infrared region.

3. METHODS

3.1 Sample preparation

(6,5)-rich CoMoCAT carbon nanotubes (Southwest NanoTechnologies, USA) were dispersed in D_2O in the presence of 0.2% (w/v) sodium dodecyl benzene sulfonate under moderate bath sonication for 60 min and vigorous sonication using a tip-type sonicator for 40 min before centrifugation at an acceleration of 130,000*g* for 4 h (ref. 19). For the experiments shown in Fig. 4, additional localized states were embedded in carbon nanotubes by atomic oxygen[32] or sp^3-defect doping[34] that reduce the band gap and exciton energy locally. Oxygen- and the sp^3-defect-doped carbon nanotubes were prepared according to procedures originally developed by Ghosh *et al.*[32] and Piao *et al.*[34], respectively, with different chemical and experimental parameter selections[33,48]. In brief, the isolated nanotube dispersion was combined with a small amount of D_2O containing dissolved ozone and left under the illumination of a desk lamp (ca. 5 mW cm^{-2}) overnight, allowing moderate atomic oxygen doping to generate sparse oxygen-derived local states. To achieve sp^3-defect doping, the nanotube dispersion was combined with a small amount of D_2O-containing dissolved organic 4-bromobenzenediazonium tetrafluoroborate salts[34] and left in the dark for at least 10 days, allowing moderate sp^3-defect doping in carbon nanotubes to generate sparse sp^3-defect-derived localized electronic states. Optical absorption spectra of these samples are presented in Supplementary Fig. 1.

3.2 Optical measurements

Optical measurements were conducted using a broadband supercontinuum white light source (Fianium, SC450) for excitation. Incident light wavelengths were selected using a monochromator for normal Stokes emission measurements and optical filters for up-conversion photoemission measurements. The typical excitation band width was 10 nm for spectroscopic measurements and any unwanted stray light was carefully removed using multiple optical filters. Normal photoluminescence measurements were performed using monochromated light at a wavelength of 568 nm (2.18 eV) that matches the E_{22} optical resonance in (6,5) carbon nanotubes for photoexcitation. The optical image showing up-conversion photoemission phenomena (Fig. 1) was taken using an electron multiplying charge coupled device (EMCCD) camera (Princeton Instruments, ProEM). The microscopic up-conversion photoemission imaging of an individual nanotube shown in Fig. 1b was performed using an inverted microscope and a 1.4 NA, × 100 objective under epi-illumination of the near-infrared incident light (1,075–1,125 nm, approximate power density of 4 kW cm^{-2}). For the microscopic imaging, the individually dispersed nanotubes were immobilized using an agarose gel matrix[37]. During spectral measurements shown in Figs 2,3, 4, the nanotube dispersion was set in a quartz cell with optical path of 1 mm. Incident light was focused on the sample in the quartz cell as a 4–6-μm spot using an achromatic lens, and light emitted from the samples was collected by an achromatic lens and transferred to a confocal optical set-up. The near-infrared photoemission was recorded using a liquid nitrogen cooled Si-CCD or an InGaAs array detector. Detection sensitivity variations were corrected using a standard light source (Ocean Optics, LS-1). The spectral resolution typically equalled 2 nm. We used spectral-integrated intensity for the analyses of the photoemission intensity as functions of excitation power density or temperature. The nanotube dispersion temperature was directly measured in a quartz cell using a dipped thin thermocouple. For the measurements of temperature dependence, nanotubes in the quartz cell were gradually cooled using iced water or allowed to warm up by natural convection. Optical spectra were acquired at temperatures ranging from 280 K to room temperature (ca. 300 K). The temperature typically varied by 1.5 K during each spectral measurement.

REFERENCES

1. Stokes, G. G. On the change of refrangibility of light. *Philos. Trans. R. Soc. Lond.* **142**,463–562 (1852).

2. Auzel, F. Upconversion and anti-stokes processes with f and d ions in solids. *Chem. Rev.***104**, 139–174 (2004).

3. Baluschev, S. *et al.* Up-conversion fluorescence: noncoherent excitation by sunlight. *Phys. Rev. Lett.* **97**, 143903 (2006).

4. Zhao, J., Ji, S. & Guo, H. Triplet–triplet annihilation based upconversion: from triplet sensitizers and triplet acceptors to upconversion quantum yields. *RSC Adv.* **1**, 937–950(2011).

5. Joly, A. G., Chen, W., McCready, D. E., Malm, J.-O. & Bovin, J.-O. Upconversion luminescence of cdte nanoparticles. *Phys. Rev. B* **71**, 165304 (2005).

6. Sheik-Bahae, M. & Epstein, R. I. Optical refrigeration. *Nat. Photon.* **1**, 693–699 (2007).

7. Rakovich, Y. P. & Donegan, J. F. in *Semiconductor Nanocrystal Quantum Dots* Ch. 9 (ed. Andrey L. Rogach 257–275Springer Vienna (2008).

8. Wang, F. *et al.* Tuning upconversion through energy migration in core–shell nanoparticles.*Nat. Mater.* **10**, 968–973 (2011).

9. Gray, V., Dzebo, D., Abrahamsson, M., Albinsson, B. & Moth-Poulsen, K. Triplet-triplet annihilation photon-upconversion: towards solar energy applications. *Phys. Chem. Chem. Phys.* **16**, 10345–10352 (2014).

10. Xu, C. T. *et al.* Autofluorescence insensitive imaging using upconverting nanocrystals in scattering media. *Appl. Phys. Lett.* **93**, 171103 (2008).

11. Vinegoni, C. *et al.* Transillumination fluorescence imaging in mice using biocompatible upconverting nanoparticles. *Opt. Lett.* **34**, 2566–2568 (2009).

12. Kotov, N. Bioimaging: The only way is up. *Nat. Mater.* **10**, 903–904 (2011).

13. Dou, Q. Q., Guo, H. C. & Ye, E. Near-infrared upconversion nanoparticles for bio-applications. *Mater. Sci. Eng. C Mater. Biol. Appl.* **45**, 635–643 (2014).

14. Downing, E., Hesselink, L., Ralston, J. & Macfarlane, R. A three-color, solid-state, three-dimensional display. *Science* **273**, 1185–1189 (1996).

15. He, G. S., Markowicz, P. P., Lin, T.-C. & Prasad, P. N. Observation of stimulated emission by direct three-photon excitation. *Nature* **415**, 767–770 (2002).

16. Epstein, R. I., Buchwald, M. I., Edwards, B. C., Gosnell, T. R. & Mungan, C. E.Observation of laser-induced fluorescent cooling of a solid. *Nature* **377**, 500–503 (1995).

17. Iijima, S. & Ichihashi, T. Single-shell carbon nanotubes of 1-nm diameter. *Nature* **363**,603–605 (1993).

18. Saito, R., Fujita, M., Dresselhaus, G. & Dresselhaus, M. S. Electronic structure of chiral graphene tubules. *Appl. Phys. Lett.* **60**, 2204–2206 (1992).

19. O'Connell, M. J. *et al.* Band gap fluorescence from individual single-walled carbon nanotubes. *Science* **297**, 593–596 (2002).

20. Ando, T. Excitons in carbon nanotubes. *J. Phys. Soc. Jpn* **66**, 1066–1073 (1997).

21. Wang, F., Dukovic, G., Brus, L. E. & Heinz, T. F. The optical resonances in carbon nanotubes arise from excitons. *Science* **308**, 838–841 (2005).

22. Maultzsch, J. *et al.* Exciton binding energies in carbon nanotubes from two-photon photoluminescence. *Phys. Rev. B* **72**, 241402 (R) (2005).

23. Bachilo, S. M. *et al.* Structure-assigned optical spectra of single-walled carbon nanotubes.*Science* **298**, 2361–2366 (2002).

24. Avouris, P., Freitag, M. & Perebeinos, V. Carbon-nanotube photonics and optoelectronics.*Nat. Photon.* **2**, 341–350 (2008).

25. Welsher, K. *et al.* A route to brightly fluorescent carbon nanotubes for near-infrared imaging in mice. *Nat. Nanotechnol.* **4**, 773–780 (2009).

26. Miyauchi, Y. Photoluminescence studies on exciton photophysics in carbon nanotubes. *J. Mater. Chem. C* **1**, 6499–6521 (2013).

27. Kruss, S. *et al.* Carbon nanotubes as optical biomedical sensors. *Adv. Drug Deliv. Rev.* **65**,1933–1950 (2013).

28. Ma, Y.-Z., Valkunas, L., Dexheimer, S. L., Bachilo, S. M. & Fleming, G. R. Femtosecond spectroscopy of optical excitations in single-walled carbon nanotubes: evidence for exciton-exciton annihilation. *Phys. Rev. Lett.* **94**, 157402 (2005).

29. Dresselhaus, M. S. & Eklund, P. C. Phonons in carbon nanotubes. *Adv. Phys.* **49**,705–814 (2000).

30. Stich, D. *et al.* Triplet-triplet exciton dynamics in single-walled carbon nanotubes. *Nat. Photon.* **8**, 139–144 (2014).

31. Crochet, J. J., Duque, J. G., Werner, J. H. & Doorn, S. K. Photoluminescence imaging of electronic-impurity-induced exciton quenching in single-walled carbon nanotubes. *Nat. Nanotechnol.* **7**, 126–132 (2012).

32. Ghosh, S., Bachilo, S. M., Simonette, R. A., Beckingham, K. M. & Weisman, R. B.Oxygen doping modifies near-infrared band gaps in fluorescent single-walled carbon nanotubes. *Science* **330**, 1656–1659 (2010).

33. Miyauchi, Y. *et al.* Brightening of excitons in carbon nanotubes on dimensionality modification. *Nat. Photon.* **7**, 715–719 (2013).

34. Piao, Y. *et al.* Brightening of carbon nanotube photoluminescence through the incorporation of sp3 defects. *Nat. Chem.* **5**, 840–845 (2013).

35. Ma, X. *et al.* Electronic structure and chemical nature of oxygen dopant states in carbon nanotubes. *ACS Nano* **8**, 10782–10789 (2014).

36. Kilina, S., Ramirez, J. & Tretiak, S. Brightening of the lowest exciton in carbon nanotubes via chemical functionalization. *Nano Lett.* **12**, 2306–2312 (2012).

37. Cognet, L. *et al.* Stepwise quenching of exciton fluorescence in carbon nanotubes by single-molecule reactions. *Science* **316**, 1465–1468 (2007).

38. Matsuda, K., Inoue, T., Murakami, Y., Maruyama, S. & Kanemitsu, Y. Exciton dephasing and multiexciton recombinations in a single carbon nanotube. *Phys. Rev. B* **77**, 033406(2008).

39. Crochet, J. J. *et al.* Disorder limited exciton transport in colloidal single-wall carbon nanotubes. *Nano Lett.* **12**, 5091–5096 (2012).

40. Rajan, A., Strano, M. S., Heller, D. A., Hertel, T. & Schulten, K. Length-dependent optical effects in single walled carbon nanotubes. *J. Phys. Chem. B* **112**, 6211–6213 (2008).

41. Moritsubo, S. *et al.* Exciton diffusion in air-suspended single-walled carbon nanotubes.*Phys. Rev. Lett.* **104**, 247402 (2010).

42. Hertel, T., Himmelein, S., Ackermann, T., Stich, D. & Crochet, J. Diffusion limited photoluminescence quantum yields in 1-D semiconductors: single-wall carbon nanotubes.*ACS Nano* **4**, 7161–7168 (2010).

43. Miyauchi, Y., Matsuda, K., Yamamoto, Y., Nakashima, N. & Kanemitsu, Y. Length-dependent photoluminescence lifetimes in single-walled carbon nanotubes. *J. Phys. Chem. C* **114**, 12905–12908 (2010).

44. Harrah, D. M. & Swan, A. K. The role of length and defects on optical quantum efficiency and exciton decay dynamics in single-walled carbon nanotubes. *ACS Nano* **5**, 647–655(2011).

45. Perebeinos, V. & Avouris, P. Phonon and electronic nonradiative decay mechanisms of excitons in carbon nanotubes. *Phys. Rev. Lett.* **101**, 057401 (2008).

46. Galland, C., Högele, A., Türeci, H. E. & Imamoğlu, A. Non-markovian decoherence of localized nanotube excitons by acoustic phonons. *Phys. Rev. Lett.* **101**, 067402 (2008).

47. Welsher, K., Sherlock, S. P. & Dai, H. Deep-tissue anatomical imaging of mice using carbon nanotube fluorophores in the second near-infrared window. *Proc. Natl Acad. Sci. USA* **108**, 8943–8948 (2011).

48. Iwamura, M. *et al.* Nonlinear photoluminescence spectroscopy of carbon nanotubes with localized exciton states. *ACS Nano* **8**, 11254–11260 (2014).

7

THE NECESSARY LENGTH OF CARBON NANOTUBES REQUIRED TO OPTIMIZE SOLAR CELLS

Majid Vaezzadeh, Mohammad Reza Saeedi, Tirdad Barghi and Mohammad Reza Sadeghi

K.N. Toosi University, Department of Physics, 41, Shahid Kavian St., 15418-49611 Tehran, Iran

ABSTRACT

Background

In recent years scientists have been trying both to increase the efficiency of solar cells, whilst at the same time reducing dimensions and costs. Increases in efficiency have been brought about by implanting carbon nanotubes onto the surface of solar cells in order to reduce the reflection of sunrays, as well as through the insertion of polymeric arrays into the intrinsic layer for charge separation.

Results

The experimental results show power rising linearly for intrinsic layer thicknesses between 0–50 nm. Wider thicknesses increase the possibility of recombination of electrons and holes, leading to perturbation of the linear behaviour of output power. This effect is studied and formulated as a function of thickness. Recognition of the critical intrinsic layer thickness can permit one to determine the length of carbon nanotube necessary for optimizing solar cells.

Conclusion

In this study the behaviour of output power as a function of intrinsic layer thicknesses has been described physically and also simulated. In addition, the implantation of carbon nanotubes into the intrinsic layer and the necessary nanotube length required to optimize solar cells have been suggested.

1. BACKGROUND

Developing inexpensive and renewable energy sources is one of the most important scientific and technological challenges of our time. Solar energy is an inexhaustible energy source that can be harnessed to meet our growing future energy needs. However, traditional photovoltaic (solar-to-electric conversion) technology is too expensive to be the suitable alternative for fossil fuels or even other competing renewable energy sources. A significant leap in the scientific and technological progress of renewable energy sources will be required to displace proven, but unsustainable energy production methods.

Nanotechnology is driving new interesting developments in photovoltaic technology. Advances in organic synthesis and characterization techniques allow us to coax a photocurrent from organic, 'soft' molecules in a process that mimics photosynthesis in plants, thus potentially opening up the way to cheap, ubiquitous solar cells. Power production resulting in zero greenhouse gas emissions is economically and environmentally desirable. Direct photovoltaic conversion of sunlight into electricity is therefore the highly attractive alternative to unsustainable energy sources such as fossil fuels.

Although silicon solar cells have gained a considerable market share and commercial success, high production and up-front installation costs still limit their commercial viability. In this study, we explore the use of low-cost advanced materials for photovoltaic energy production and the mechanism of photovoltaic action in a new class of solar cell, that is, the heterojunction photovoltaic. These are constructed from a thin film of cheap composite material, a mixture of carbon nanotubes and conductive polymer [1].

2. RESULTS

We now aim to address the issues that determine the thickness of the absorber (or "intrinsic") layer. Figure 1 illustrates a computational calculation that shows how the output power of an a-Si-based *pin*cell varies with intrinsic layer thickness. The curves differ in the specified absorption coefficients for a monochromatic illumination using varying photon energies. All curves were calculated for the same photon flux. Such illumination conditions might be achieved by experiment using a laser whose photon energy can be tuned from 1.8 to 2.3 eV, however the presence of sunlight, of course, presents a much more complex situation.

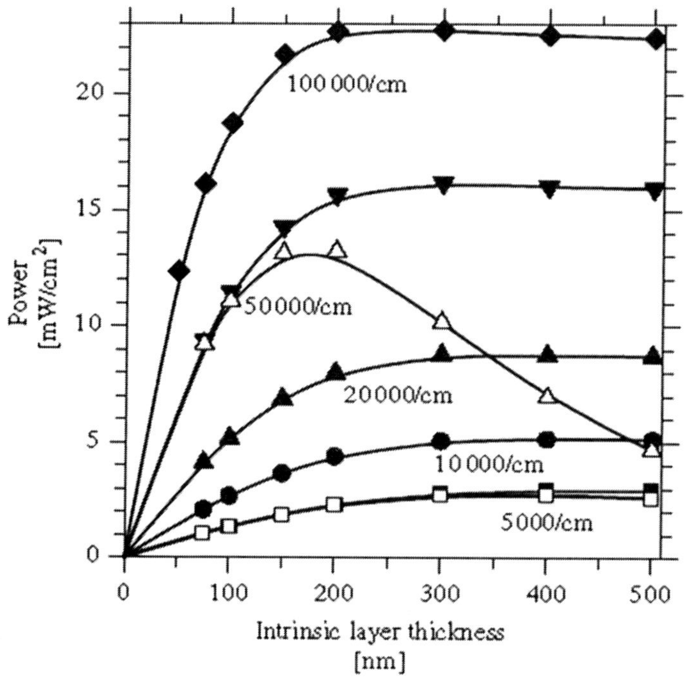

Figure 1. Computer calculation of the power output from a *pin* solar cell as a function of intrinsic layer thickness. The differing curves indicate results for monochromatic illumination with absorption coefficients from 5000/cm to 100 000/cm; for typical a-Si: H, this range corresponds to a photon energy range from 1.8 to 2.5 eV. Solid symbols indicate illumination through the *p*-layer and open symbols indicate illumination through the *n*-layer. Incident photon flux = 2 × 1017/cm2s; no back reflector. The data are obtained from real experiments, with the curves calculated using methods outlined in the references [3-8].

We will first discuss the results for illumination through the *p*-layer (solid symbols in the figure). For intrinsic layers that are sufficiently thin, the power is proportional to the number of photons absorbed (i.e. to the product of the thickness, *d* and the absorption coefficient, η). Within this limit the fill factors have nearly ideal values around 0.8. As the thickness of the cell increases, the power saturates. In figure 1 the first plot (η = 100 000/cm – corresponding photon energy of about 2.3 eV) shows the behaviour for strongly absorbed illumination. Power saturation occurs for thicknesses greater than 100 nm, that is, typical distance at which photons are absorbed. Since thicker cells do not absorb much additional light, going beyond this length means power stops increasing [2].

2.1 Model and objective

Different models have been proposed to simulate the experimental results [3, 4, 5, 6, 7, 8]. Figure 1shows the behaviour of power as a function of intrinsic layer thickness. Several experiments have been carried out to demonstrate the effect of layer thickness on output power.

As can be observed in figure 1, power grows linearly and becomes saturated after a distinct value of layer thickness. If we neglect recombination of electrons and holes, the output power can be expressed as:

$$P_{\alpha} = \alpha w, \tag{1}$$

where, α is a coefficient and can be calculated for each curve, w is the intrinsic layer thickness, with P_{α} termed the ideal power (linear form).

Because the recombination phenomena begin at a certain intrinsic layer thickness then it is useful to plot P_{α}-P_{exp} as a function of the intrinsic layer thickness. Figure 2 shows the difference between the experimental power and P_{α} as a function of the intrinsic layer for P_1 type with α = 0.24 and η = 100 000/cm (absorption coefficients).

On the basis of figure 2 we can produce an equation that contains the recombination phenomena as:

$$P = P_{\alpha} - \beta(w - w_c)^n, \quad (2)$$

where β is the recombination coefficient. This equation is plotted in figure 3 with β = 0.02 and n = 1.41.

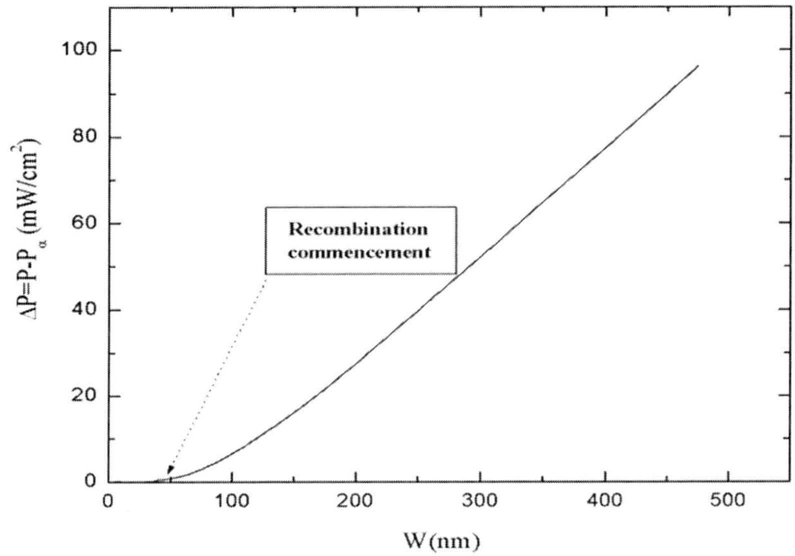

Figure 2. The difference between ideal (linear form) and experimental power as a function of the intrinsic layer thickness for P_1 type.

Figure 3 shows the ideal (linear form), theoretical and experimental powers as a function of the intrinsic layer thickness for P_1 type. The recombination phenomena start at W_c = 40 nm, which we call the critical intrinsic layer.

From figure 3 it can be seen that the theoretical power values between 0–150 nm almost match the experimental results. However after 150 nm the agreement between the fits is poor. When the difference between the intrinsic layer thickness and carrier current cross section is small, the two-dimensional recombination phenomena are expected to dominate (from 40 nm up to 150 nm for P_1type). Whereas, when the difference between the intrinsic layer thickness and carrier current cross section is large (after 150 nm for P_1 type) the three-dimensional recombination phenomena dominate. On increasing the length of intrinsic layer, the three-dimensional recombination becomes dominant and exhibits linear behaviour.

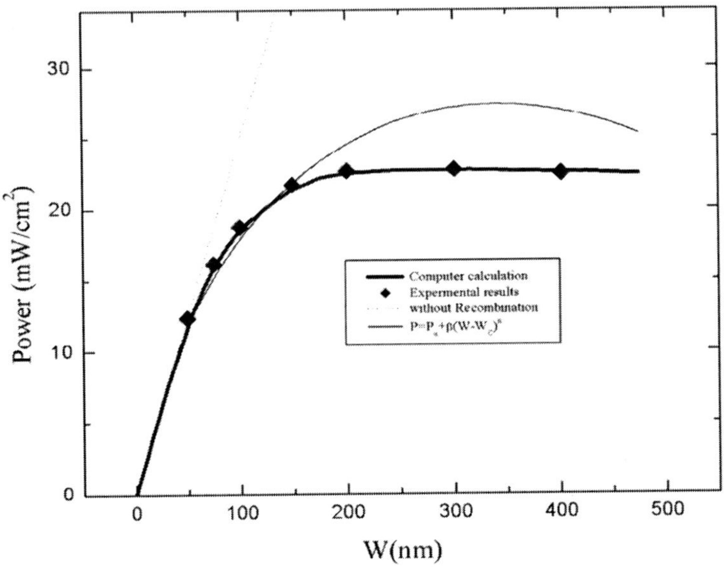

Figure 3. Ideal, theoretical and experimental power as a function of the intrinsic layer thickness for P_1 type.

Therefore, the theoretical model plot can be defined by three regions that is, linear (0–40 nm), two-dimensional and three-dimensional recombination phenomena:

$$P = \alpha w - \beta(w - w_c)^n + (a - \alpha' w), \qquad (3)$$

where a and α' are the three dimensional recombination coefficients.

Because of interference between two-dimensional and three-dimensional recombination terms in equation 3, this equation is not appropriate for simulation. Consequently we need to modify equation 3 thus:

$$P_{th} = \alpha w - \beta(|w - w_c|)^n X(w) + (a - \alpha' w)Y(w), \tag{4}$$

where X (w) is the creation and destruction function and Y (w) is the creation function.

We shall now define the function, starting fist with the destruction function,

$$F(w) = \frac{1}{1 + e^{\gamma(\frac{w}{w_C} - 1)}} \tag{5}$$

where γ is function sharpness parameter.

Figure 4 shows the plot of the destruction function against different sharpness parameters.

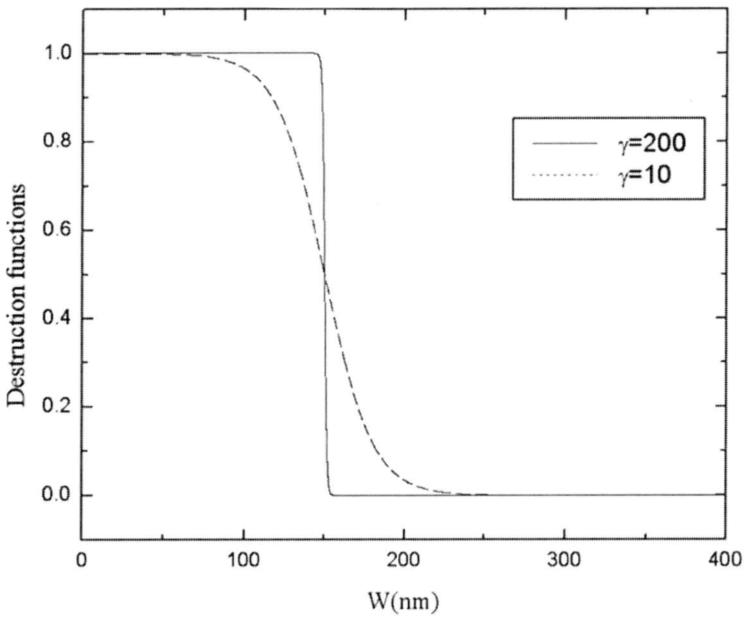

Figure 4. Destruction functions as a function of different sharpness parameters.

Second, the creation function can be defined as:

$$Y(w) = 1 - F(w) = \frac{1}{1 + e^{-\gamma(\frac{w}{w_C} - 1)}} \tag{6}$$

Lastly, we can express the creation and destruction function X(w) as:

$$X(w) = \frac{1}{1 + e^{\gamma\left[\frac{w}{(w_C + \delta w)} - 1\right]}} \times \frac{1}{1 + e^{-\gamma\left[\frac{w}{(w_C)} - 1\right]}} \tag{7}$$

The behaviour of equation 7 is shown in figure 5. We can calculate the data for the theoretical power plot by using the functions X (w), Y (w) and equation. 4. The data plotted for the calculation of theoretical power with a function sharpness parameter equal to 5 are shown in table 1. It is important to note that the recombination starts at w_c. A plot using the data outlined in table 1 is shown in figure 6.

The necessary length of carbon nanotube for an optimized solar cell

Because of the symmetry and unique electronic structure of graphene, the nanotube's structure strongly affects its electrical properties. For a given (*n*, *m*) nanotube, if *n* – *m* is a multiple of 3, then the nanotube is metallic; such is the case for all armchair (*n* = *m*) nanotubes. In terms of metallic behaviour, carbon nanotubes can have conductivity up to eight times that of copper, and can carry a current density comparable to that of any known conventional metallic wire; this makes carbon nanotubes potential candidates as nanoscale wires [9]. It is consequently possible to reduce the recombination phenomena by using the carbon nanotubes in the intrinsic layer. The presence of an aligned doped carbon nanotube (armchair, superconducting) in the intrinsic layer helps maintain linear power generation behaviour, in addition to making solar cells with higher output powers.

It is necessary to know the proper length of doped carbon nanotubes as a function of intrinsic layer thickness. The non-linearity of power starts at w_c, which for P_1 type, for example, starts at 40 nm. So the necessary length of carbon nanotube for optimizing a solar cell, as a function of the intrinsic layer, can be expressed as:

$$L_{nanotube} = w - w_C \tag{8}$$

Table 1. Data plotted for the calculation of theoretical power (γ = 5).

type	α	β	w_c	n	δw	a	α'
P1 (100 000/cm)	0. 24	0.02	40	1.41	120	22.71	0.2405
P2 (50 000/cm)	0.124	0.0011	55	1.85	140	17.3	0.1265
P3 (20 000/cm)	0.053	0.0004	70	1.85	170	8.96	0.053
P4 (10 000/cm)	0.024	0.000065	97	2.0105	230	6	0.0249
P5 (5000/cm)	0.0116	0.0000035	110	2.31	260	2.1	0.01

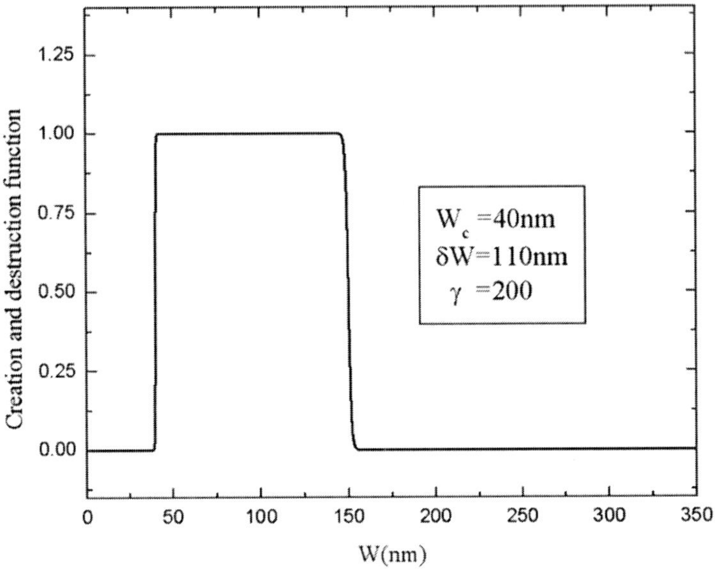

Figure 5. Creation and destruction function at different sharpness parameters.

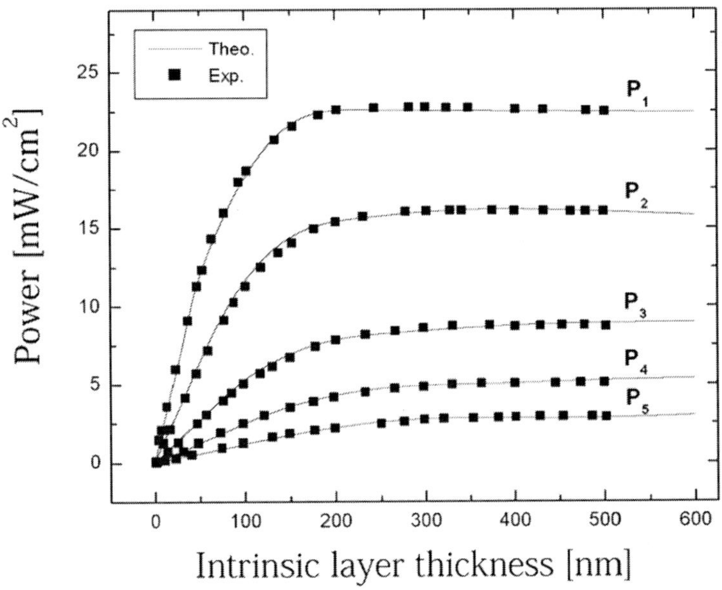

Figure 6. Simulated plot of power as a function of intrinsic layer thickness.

It is important to note that the volume occupied by carbon nanotubes in the intrinsic layer affects only the linearity coefficient.

3. CONCLUSION

When considering the behaviour of power as a function of intrinsic layer thickness, the recombination of electrons and holes plays an important role in reducing the power. Eliminating recombination is useful in maintaining the linear behaviour of power and in producing solar cells with high powers. If the intrinsic layer thickness is comparable to the carrier current cross section, the two dimensional recombination phenomena are dominant; otherwise the three dimensional recombination phenomena would dominate. In this study, power as a function of the intrinsic layer thickness was simulated and plotted, thus allowing us to express the necessary length of carbon nanotubes required for optimized solar cells as a function of the intrinsic layer thickness.

REFERENCES

1. Motta N, Waclawik ER, Goh R, Bell JM: Nanotube-polymer Solar Cells – an Alternative to Silicon. Bollettino della Comunità Scientifica in Australasia. 2005, 15-20.

2. Luque A, Hegedus S: John Wiley & Sons Handbook of Photovoltaic Science and Engineering. 2003, Ltd ISBN: 0-471-49196-9

3. Davis Mark, Hunter Fanney A, Dougherty Brian: Evaluating building integrated photovoltaic performance models. National Institute of Standards and Technology, Gaithersburg, MD. 20899, -8632

4. Dianove M, Kazansky PG, Stepanovd Yu: Photovoltaic model of photoinduced second-harmonic generation in optical fibers. Acad Sci USSR. General physics inst., dep. fiber optics, Moscow 117942, URSS (Sov. lightwave commun.)

5. De Blas MA, Torres JL, Prieto E, Garca A: Selecting a suitable model for characterizing photovoltaic devices. Renewable Energy. 2002, 25 (3): 371-380(10). 10.1016/S0960-1481(01)00056-8.

6. Cristofari C, Poggi P, Notton G, Muselli M: Thermal Modelling of a Photovoltaic Module. Proceeding Modelling, Simulation, and Optimization. 2006, 507:

7. Hunter Fanney A, Dougherty Brian, Davis Mark: Short-Term Characterization of Building Integrated Photovoltaic Panels. National Institute of Standards and Technology. 2002, 100 Bureau Drive, MS 8632, Gaithersburg, MD 20899-8632

8. Munoz-Perez MA, Rodriguez-Pichardo A, Camacho Martinez F, Sidrach-de-Cardona M, Mora Lopez L: A simple model for sizing stand alone photovoltaic systems. Solar Energy Materials and Solar Cells. 55 (3): 199-214(16). 24 August 1998

9. Appenzeller J, Martel R, Derycke V, Radosavljevic M, Wind S, Neumayer D, Avouris P: Carbon nanotubes as potential building blocks for future nanoelectronics. Microelectronic Engineering. 2002, 64 (1-4): 391-7. 10.1016/S0167-9317(02)00813-4.

8

GENERAL SYNTHESIS OF COMPLEX NANOTUBES BY GRADIENT ELECTROSPINNING AND CONTROLLED PYROLYSIS

Chaojiang Niu[1], Jiashen Meng[1], Xuanpeng Wang[1], Chunhua Han[1], Mengyu Yan[1], Kangning Zhao[1], Xiaoming Xu[1], Wenhao Ren[1], Yunlong Zhao[1,2], Lin Xu[2], Qingjie Zhang[1], Dongyuan Zhao[1] & Liqiang Mai[1]

[1]State Key Laboratory of Advanced Technology for Materials Synthesis and Processing, Wuhan University of Technology, Wuhan 430070, China.

[2]Department of Chemistry and Chemical Biology, Harvard University, Cambridge, Massachusetts 02138, USA

Nanowires and nanotubes have been the focus of considerable efforts in energy storage and solar energy conversion because of their unique properties. However, owing to the limitations of synthetic methods, most inorganic nanotubes, especially for multi-element oxides and binary-metal oxides, have been rarely fabricated. Here we design a gradient electrospinning and controlled pyrolysis method to synthesize various controllable 1D nanostructures, including mesoporous nanotubes, pea-like nanotubes and continuous nanowires. The key point of this method is the gradient distribution of low-/middle-/high-molecular-weight poly(vinyl alcohol) during the electrospinning process. This simple technique is extended to various inorganic multi-element oxides, binary-metal oxides and single-metal oxides. Among them, $Li_3V_2(PO_4)_3$, $Na_{0.7}Fe_{0.7}Mn_{0.3}O_2$ and Co_3O_4 mesoporous nanotubes exhibit ultrastable electrochemical performance when used in lithium-ion batteries, sodium-ion batteries and supercapacitors, respectively.

We believe that a wide range of new materials available from our composition gradient electrospinning and pyrolysis methodology may lead to further developments in research on 1D systems.

One-dimensional (1D) nanostructures, including nanowires and nanotubes, have been a focus of nanoscience and nanotechnology, due to the unique low-dimensional properties[1, 2, 3, 4, 5, 6]. Various successful fabrication techniques, such as chemical/physical vapour deposition, hydrothermal, template based, electrochemical etching/deposition, laser ablation and electrospinning methods[7, 8, 9, 10, 11, 12, 13, 14, 15], have been developed for specific materials. However, owing to the restriction of applicable objects of each synthetic method and the difference of crystal growth orientation of different substances[16, 17, 18], 1D nanostructures are only achievable for some specific materials. Most inorganic nanotubes, especially for multi-element oxides and binary-metal oxides, have been scarcely obtained, which greatly restricts their further developments. Therefore, a universal technique is required, which can be used to fabricate nanotubes, as well as nanowires, for various inorganic materials with taking no account of the limitation of crystal orientation.

Electrospinning techniques have been studied for the fabrication of conductive polymer nanowires and a part of inorganic material nanowires[19, 20, 21]. This method along with different post-treatments has been applied to synthesize some interesting surface multilevel structures (branched nanowires and necklace-like nanowires) and inner multilevel structures (core/shell nanowires and multichannel microtubes)[22, 23, 24, 25, 26]. However, most of these structures have been confined to solid nanowires rather than nanotubes. Xia and co-workers have proposed a melt coaxial electrospinning method, with a coaxial spinneret, for fabricating core-shell nanowires and metal nanotubes[27, 28, 29]. Jiang and co-workers have developed a multifluidic compound-jet electrospinning technique for fabricating biomimic multichannel microtubes[30]. Nevertheless, it has been rarely proposed to electrospin nanotubes for different kinds of inorganic oxides, especially for multi-element oxides and binary-metal oxides, with abundant materials diversity, low cost, good repeatability and high yield, which seriously limit their further applications.

Here we design a universal gradient electrospinning followed by controlled pyrolysis methodology to synthesize various types of mesoporous nanotubes and pea-like nanotubes (Fig. 1), including multi-element oxides, binary-metal oxides and single-metal oxides. This strategy is achieved through electrospinning with one ordinary syringe needle while modulating low-, middle- and high-molecular-weight poly(vinyl alcohol) (PVA) in the precursor. In this way, different nanotubes are obtained using controllable heat treatments. The resulting mesoporous nanotubes are composed of ultrathin carbon nanotubes (~5 nm in thickness and over 10 μm in length) and small nanoparticles (approximately 5–20 nm in diameter) on the tube walls. Pea-like nanotubes are composed of outer carbon nanotubes (~20 nm in thickness) and nanoparticles (approximately 100–300 nm in diameter) in the nanotubes. These structures have larger specific surface area and higher ionic–electronic conductivity compared with traditional nanowires, which exhibits great potential in energy storage fields. Therefore, $Li_3V_2(PO_4)_3$, $Na_{0.7}Fe_{0.7}Mn_{0.3}O_2$ and Co_3O_4mesoporous nanotubes were selected as electroactive materials in lithium-ion batteries, sodium-ion batteries and supercapacitors, respectively.

1. GRADIENT ELECTROSPINNING AND CONTROLLED PYROLYSIS MECHANISM

First, the viscous homogeneous precursor solution was prepared by mixing low-, middle- and high-molecular-weight PVA (in a weight ratio of 3:2:1) and different needed inorganic materials. The precursor solution was then delivered into an ordinary metallic needle at a constant flow rate and electrospun at 20 kV. Under the strong electrostatic tension force, the low-, middle- and high-molecular-weight PVA tend to be separated into three layers instead of mixing together. The reasons are as follows: at the same electrospinning conditions, the electric field (E), flow rate (Q), electric current (I), distance (D) between the injector nozzle and the receiver, and other operating characteristics of the low-, middle- and high-molecular-weight PVA are the same. According to the two equations proposed by Baumgarten and Rutledge[31, 32] (equations. 1, 2),

$$R = c \cdot \eta^{1/2} \tag{1}$$

$$R = c(I/Q)^{-2/3}\gamma^{1/3} \tag{2}$$

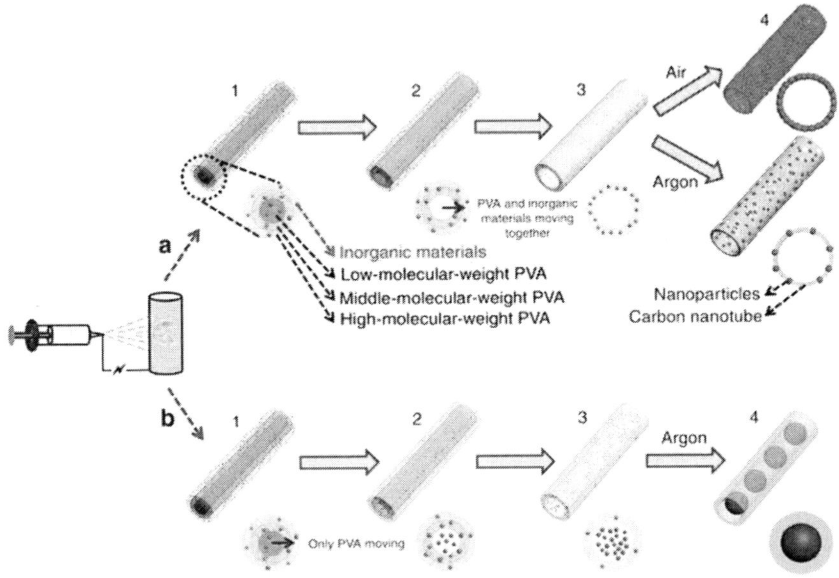

Figure 1. Schematics of the gradient electrospinning and controlled pyrolysis method. (**a**) Preparation process of mesoporous nanotubes. (1) After the electrospinning process, the low-, middle- and high-molecular-weight PVA tend to be distributed into three layers in the radial direction of composite nanowires. (2) As the temperature is slowly increased, the inner low-weight PVA first pyrolyses and moves towards the boundary of the low-/middle-weight PVA, carrying the inorganic materials. Then the middle-weight PVA pyrolyses and moves towards the high-weight PVA as well. (3) All of the preliminary pyrolysed PVA and inorganic materials converge together in the tube walls. (4) After sintering in air, all of the PVA pyrolyse and uniform mesoporous nanotubes are obtained, which are composed of tiny inorganic nanoparticles. On the other hand, after high-temperature sintering under argon, PVA carbonize, uniform mesoporous nanotubes are also obtained, which consists of inorganic nanoparticles and carbon nanotubes. The mesopores result from the decomposition of the inorganic materials and a part of PVA polymers. (**b**) Preparation process of pea-like nanotubes. (1) After the electrospinning process, the composite nanowires are directly and immediately placed into a furnace in air, which is preheated to and maintained at 300 °C. (2, 3) All of the PVA decompose at the same time and rapidly move towards the outer high-weight PVA layer without carrying the inorganic materials, leaving them in the centre. (4) After high-temperature sintering under argon, the outer PVA carbonize and the inner inorganic materials develop into nanoparticles, forming pea-like nanotubes.

where R is the terminal jet radius, η is the viscosity, c is a constant, γ is the surface tension, $Q$is the flow rate and I is the electric current. The terminal jet radius (R) is directly proportional to the square root of the viscosity ($\eta^{1/2}$) and to the cube root of the surface tension ($\gamma^{1/3}$). The ηvalues of the low-, middle- and high-molecular-weight PVA were measured as 0.0766, 0.5350 and 0.7685, dl g^{-1}, respectively, increasing gradually. The $\boldsymbol{\gamma}$ values of these three PVA were tested as 40.1, 41.6 and 51.4 mN m^{-1}, respectively (Supplementary Fig. 1a-c, g). Therefore, the high-weight PVA was distributed in the outer layer, the middle-weight PVA was located in the middle layer, and the low-weight PVA was concentrated in the centre in theory (Fig. 1a). To prove this important viewpoint, low-weight PVA was replaced by polyvinyl pyrrolidone (PVP), which has much smaller molecular weight, viscosity and surface tension (Fig. 2a, Supplementary Fig. 1g). After electrospinning low-, middle- and high-molecular-weight PVA composite polymer, the sample is solid nanowires, as shown in transmission electron microscope (TEM) image (Fig. 2b). After replacing the low-weight PVA by PVP, the sample is solid nanowires as well (Fig. 2c). Herein, PVP can be dissolved in trichloromethane ($CHCl_3$) solution, but PVA can not be dissolved in it. Therefore, the composite polymer nanowires were soaked in the $CHCl_3$ to remove PVP, getting middle-/high-molecular-weight PVA polymer nanotubes (Fig. 2d), which can clearly prove the layered distribution of low-, middle- and high-molecular-weight PVA. At the same time, the inorganic materials were homogeneously dispersed in all three layers.

After the electrospinning process, the composite nanowires were presintered in air. According to the thermogravimetric curves, the differentials of the mass loss (M) and temperature (T) ($\mathrm{d}M/\mathrm{d}T$) of the low-, middle- and high-weight PVA are −1.50, −0.95 and −0.58, respectively (Supplementary Fig. 1d,e). Consequently, the inner low-weight PVA first pyrolyzes and shrinks as the temperature is slowly increased, and moves towards the boundary of the low-/middle-weight PVA, carrying the inorganic materials simultaneously, thereby leading to the formation of nanotubes[33]. By this analogy, the middle-weight PVA then pyrolyzes and moves towards the middle-/high-weight PVA, together with the inorganic materials, thereby leading to the expansion of the inner diameter of nanotubes, as illustrated in Fig. 1a. Finally, all of the preliminarily decomposed PVA and inorganic materials converge together on the outer tubes. At last, inorganic mesoporous nanotubes can be obtained after sintering in air, which are only composed of tiny inorganic nanoparticles. Mesopores are formed via the decomposition of the inorganic materials and the partial pyrolysis of PVA. On the other hand, after high-temperature sintering under argon, all of the PVA carbonize, resulting in composite mesoporous nanotubes, which are composed of uniform inorganic nanoparticles and ultrathin carbon nanotubes.

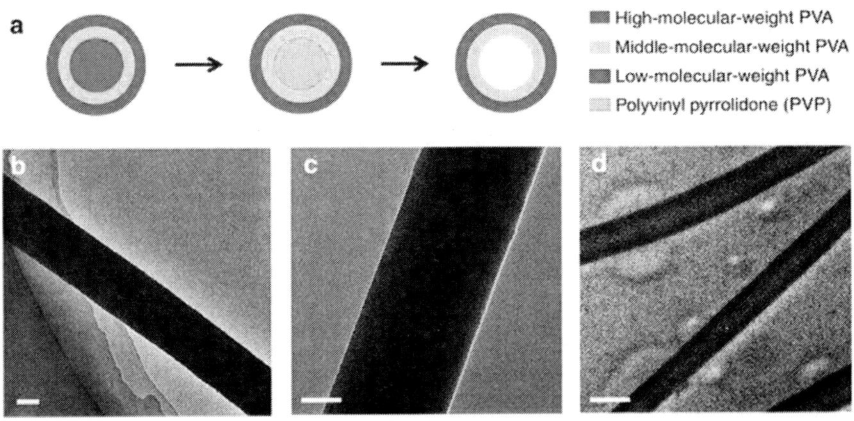

Figure 2. Schematic and characterization of the gradient distribution of low-/middle-/high-molecular-weight PVA. (**a**) Schematic of the process of replacing low-molecular-weight PVA by PVP, then removing PVP with trichloromethane, to prove the layered distribution of low-, middle- and high-molecular-weight PVA. (**b**) TEM image of low-, middle- and high-molecular-weight PVA composite polymer nanowire after electrospinning with a scale bar at 100 nm. (**c**) TEM image of PVP and middle-/high-molecular-weight PVA composite polymer nanowire after electrospinning with a scale bar at 50 nm. Low-molecular-weight PVA is replaced by PVP. (**d**) TEM image of middle-/high-molecular-weight PVA polymer nanotubes with a scale bar at 500 nm, after removing PVP in the inner center using $CHCl_3$.

For pea-like nanotubes, the preliminary electrospinning process of pea-like nanotubes is the same as that of the mesoporous nanotubes. After electrospinning, the composite nanowires were directly and immediately placed into the furnace, which was preheated to 300 °C in air. At this temperature, all of the low-/middle-/high-weight PVA simultaneously decompose (Supplementary Fig. 1d) and quickly move towards the outer high-weight PVA layer without carrying the inorganic materials in the radial direction (Fig. 1b). Thus, the inorganic materials are left *in situ* in the centre of the nanotubes. When the samples are annealed at high temperature under argon, the outer preliminary decomposed PVA carbonize and form carbon nanotubes, whereas the inner inorganic materials develop into nanoparticles, which are uniformly dispersed in the nanotubes. Eventually, pea-like nanotubes are obtained.

To confirm the mechanism of our gradient electrospinning and controlled pyrolysis method, various inorganic materials were electrospun into mesoporous nanotubes and pea-like nanotubes according to the aforementioned procedures (Fig. 3). First, multi-element oxides ($Li_3V_2(PO_4)_3$, $Na_3V_2(PO_4)_3$, $Na_{0.7}Fe_{0.7}Mn_{0.3}O_2$ and $LiNi_{1/3}Co_{1/3}Mn_{1/3}O_2$) were electrospun into uniform mesoporous nanotubes with a diameter of ~200 nm. Then, the binary-metal oxides ($LiMn_2O_4$, $LiCoO_2$, $NiCo_2O_4$ and LiV_3O_8) were electrospun into mesoporous nanotubes with a diameter of ~150 nm. For single-metal oxides (CuO, Co_3O_4, SnO_2 and MnO_2), mesoporous nanotubes with a smaller diameter of ~50 nm were fabricated. For the pea-like nanotubes, Co, $LiCoO_2$, $Li_3V_2(PO_4)_3$and $Na_{0.7}Fe_{0.7}Mn_{0.3}O_2$ were selected from the different species, the outer layer was carbon and the inner particles were different inorganic salts. Additional scanning electron microscope (SEM) images and corresponding X-ray diffraction patterns of each sample are presented as well (Supplementary Figs 2 and 3). The detailed processes are clearly illustrated in the Methods section. Each precursor solution was electrospun at a constant flow rate of ~0.1 ml h^{-1} using one ordinary syringe needle (10 ml), and the corresponding production of inorganic materials was as high as ~0.6 mmol h^{-1} and ~900 cm^2 of aluminium foil for each run (Supplementary Fig. 1f), which was a large yield. And these products can also be collected in a parallel array around the rotating wheel, to improve the packing density.[34]

2. ELECTROCHEMICAL CHARACTERIZATION

In energy-storage fields, most electrodes reported previously for batteries and supercapacitors suffer problems associated with a low conductivity, a small electrolyte/electrode surface area and self-aggregation during the charge/discharge process, leading to unsatisfactory performance, which greatly limits their applications[35, 36, 37, 38, 39, 40, 41, 42, 43, 44, 45]. One-dimensional nanomaterials have widely been investigated and applied in energy storage fields due to its unique low-dimensional properties. Remarkably, our complex nanotubes, especially mesoporous nanotubes, have the characteristics of large surface area, excellent stability and continuous carbon nanotubes with high conductivity and so on, which is expected to effectively improve the electrochemical performance of electrodes (Fig. 4). Therefore, to confirm it, $Li_3V_2(PO_4)_3$, $Na_{0.7}Fe_{0.7}Mn_{0.3}O_2$ and Co_3O_4 mesoporous nanotubes were selected and measured as typical examples of electroactive materials in lithium-ion batteries, sodium-ion batteries and supercapacitors, respectively.

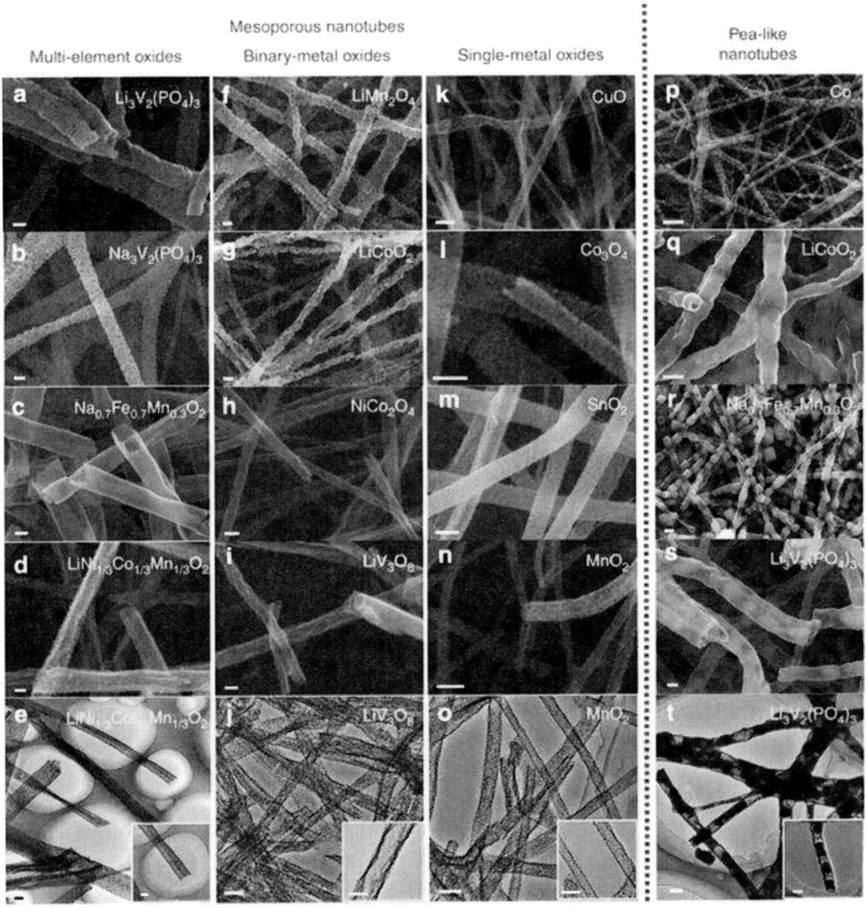

Figure 3. Expansion of the gradient electrospinning and controlled pyrolysis method. (**a–o**) SEM and TEM images of multi-element oxides ($Li_3V_2(PO_4)_3$, $Na_3V_2(PO_4)_3$, $Na_{0.7}Fe_{0.7}Mn_{0.3}O_2$and $LiNi_{1/3}Co_{1/3}Mn_{1/3}O_2$), binary-metal oxides ($LiMn_2O_4$, $LiCoO_2$, $NiCo_2O_4$ and LiV_3O_8) and single-metal oxides (CuO, Co_3O_4, SnO_2 and MnO_2) mesoporous nanotubes, respectively, scale bars, 100 nm. (**p–t**) SEM and TEM images of pea-like nanotubes (Co, $LiCoO_2$, $Li_3V_2(PO_4)_3$ and $Na_{0.7}Fe_{0.7}Mn_{0.3}O_2$) from different species with scale bars at 200 nm. The scale bars for the inset TEM images (**e**, **j**, **o**, **t**) are 100 nm.

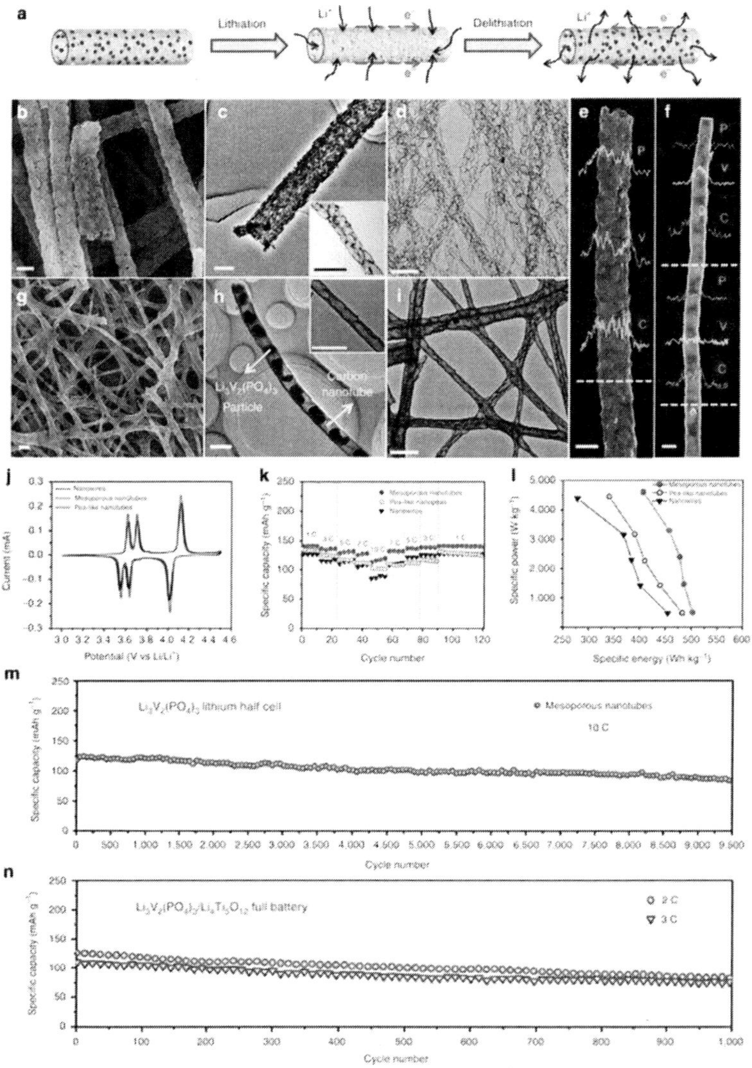

Figure 4. Characterization and electrochemical performance in lithium-ion batteries. (**a**) Schematic of the lithiation and delithiation processes of mesoporous nanotubes. (**b,c**), SEM and TEM images of $Li_3V_2(PO_4)_3$ mesoporous nanotubes with scale bar at 200 nm. (**d**) and the inset of (**c**), TEM images of ultrathin carbon nanotubes after the $Li_3V_2(PO_4)_3$ was removed using hydrogen fluoride with scale bar at 500 nm. (**e,f**) Energy-dispersive X-ray spectra (EDS) line scans of $Li_3V_2(PO_4)_3$ mesoporous nanotubes (**e**) and pea-like nanotubes (**f**) with scale bar at 200 nm, respectively. (**g,h**) SEM and TEM images of $Li_3V_2(PO_4)_3$ pea-like nanotubes with scale bar at 500 nm. (**i**) and the inset of (**h**) TEM images of carbon nanotubes after removing $Li_3V_2(PO_4)_3$ with hydrogen fluoride

with scale bar at 500 nm. (**j**) Cyclic voltammograms (CV) of the half cells collected at a sweep rate of 0.1 mV s^{-1} in the potential ranging from 3 to 4.5 V versus Li/Li$^+$. (**k,l**) Rate performance and the corresponding Ragone plots of these three cathodes measured at the rates of 1, 3, 5, 7 and 10 C, respectively. (**m**) Long cycling performance of $Li_3V_2(PO_4)_3$ mesoporous nanotubes measured at 10 C for a large number of 9,500 cycles in a lithium half cell. (**n**) Cycling performance of $Li_3V_2(PO_4)_3/Li_4Ti_5O_{12}$ lithium-ion full batteries measured at 2 and 3 C for 1,000 cycles (1 C equals to 133 mA g^{-1}).

First, the $Li_3V_2(PO_4)_3$ mesoporous nanotubes were further characterized, measured and compared with the $Li_3V_2(PO_4)_3$ pea-like nanotubes and solid nanowires (Supplementary Figs. 4-7). The $Li_3V_2(PO_4)_3$ mesoporous nanotubes consist of ultrathin, continuous, mesoporous carbon nanotubes (~200 nm in diameter) and $Li_3V_2(PO_4)_3$ nanoparticles (approximately 20–50 nm, uniformly dispersed on the tubes) (Fig. 4b–d), which is clearly confirmed by removal of the $Li_3V_2(PO_4)_3$ with hydrogen fluoride. Moreover, the mesopore size can also be tuned by modulating the sintering temperature (Supplementary Fig. 5). In contrast, the $Li_3V_2(PO_4)_3$ pea-like nanotubes are composed of outer thick carbon nanotubes (~200 nm in diameter and ~20 nm in wall thickness) and $Li_3V_2(PO_4)_3$ nanoparticles in the nanotubes (~200 nm in diameter), which uniformly separate from each other (Fig. 4g–i). The EDS line scans demonstrate the homogeneous distribution of V, P and C in mesoporous nanotubes, and display the difference between the hollow and solid parts in pea-like nanotubes (Fig. 4e,f). The carbon mass content of $Li_3V_2(PO_4)_3$ mesoporous nanotubes is only 7%, which is much smaller compared with pea-like nanotubes (17%) and nanowires (12%; Supplementary Fig. 4). $Li_3V_2(PO_4)_3$ mesoporous nanotubes exhibit higher capacity and better rate recovery (100%) than pea-like nanotubes (94%) and nanowires (98%) when tested at various rates ranging from 1, 3, 5, 7 to 10 C, demonstrating prominent rate performance (Fig. 4j–l). Particularly, $Li_3V_2(PO_4)_3$ mesoporous nanotubes can operate stably for as long as 9,500 cycles at the high rate of 10 C. The 9,500th discharge capacity is 86 mAh g^{-1}, which corresponds to a capacity retention of 80% and to a capacity fading of 0.0024% per cycle (Fig. 4m). However, the capacity retentions of $Li_3V_2(PO_4)_3$ pea-like nanotubes and solid nanowires are only 71% and 68%, respectively, after being cycled only 1,100 times at 10 C (Supplementary Fig. 6a). Then further electrochemical measurements of $Li_3V_2(PO_4)_3$ mesoporous nanotubes were implemented. When $Li_3V_2(PO_4)_3$ mesoporous nanotubes were measured at different temperatures of -20, 20 and

60 °C at a rate of 5 C, the corresponding initial capacities were 103, 131 and 136 mAh g^{-1}, respectively (Supplementary Fig. 6c,d). And the coulombic efficiency of $Li_3V_2(PO_4)_3$ mesoporous nanotubes can be kept on ~100% when tested at 1 C (133 mAh g^{-1}) for 500 cycles (Supplementary Fig. 6e). In the meantime, we assembled $Li_3V_2(PO_4)_3/Li_4Ti_5O_{12}$ lithium-ion full batteries. $Li_4Ti_5O_{12}$ was selected as the anode because of its stable charge/discharge plateau (Supplementary Fig. 7a–c). This full battery delivers capacities of 118, 93, 80 and 74 mAh g^{-1} at the rates of 2, 3, 4 and 5 C, respectively, and recovers 96% of its initial capacity when the rate is decreased back to 2 C (Supplementary Fig. 7d). Notably, the $Li_3V_2(PO_4)_3/Li_4Ti_5O_{12}$ battery is stably charged/discharged for over 1,000 cycles at rates of 2 and 3 C, with a capacity retention of 73% and 75%, respectively (Fig. 4n), demonstrating an excellent cycling performance and great commercial potential.

The superior performance is explained as follows: (1) $Li_3V_2(PO_4)_3$ mesoporous nanotubes contain ultrathin continuous carbon nanotubes. The charge transfer resistance (Rct) of $Li_3V_2(PO_4)_3$mesoporous nanotubes is measured as 56 Ω, much smaller than that of nanowires (115 Ω), which can obviously improve the electronic conductivity[36] (Supplementary Fig. 6b). (2) $Li_3V_2(PO_4)_3$mesoporous nanotubes exhibit the largest specific surface area (115 $m^2 g^{-1}$) among mesoporous nanotubes, pea-like nanotubes (33 $m^2 g^{-1}$) and nanowires (15 $m^2 g^{-1}$), which is beneficial to increase the electrode–electrolyte contact area and the active sites[39, 40] (Supplementary Fig. 4). (3) These nanotubes have abundant mesopores on the tube walls, which can effectively buffer the stress induced during the charge/discharge process, and greatly improve the structural stability[40] (Fig. 4a).

The $Na_{0.7}Fe_{0.7}Mn_{0.3}O_2$ mesoporous nanotubes are also composed of ultrathin carbon nanotubes (~200 nm in diameter) and $Na_{0.7}Fe_{0.7}Mn_{0.3}O_2$ nanoparticles (~10 nm) on the tubes (Fig. 5a). And the X-ray diffraction pattern, ICP measurement and high-resolution TEM demonstrate the rhombohedral crystal structure of $Na_{0.7}Fe_{0.7}Mn_{0.3}O_2$ (Supplementary Fig. 8). Stable voltage plateaus are observed when measured at the current densities of 100, 200, 300 and 500 mA g^{-1}, respectively, corresponding to one pair of well-defined anodic (3.9 V) and cathodic (3.6 V) peaks (Fig. 5b). When $Na_{0.7}Fe_{0.7}Mn_{0.3}O_2$ mesoporous nanotubes are tested at a low density of 100 mA g^{-1} in the potential range 3–4.5 V, 90% of the initial capacity (109 mAh g^{-1}) is retained after 1,000 cycles (Fig. 5c). When measured at a high current density of 500 mA g^{-1}, 70% of the initial capacity (82 mAh g^{-1}) is maintained after cycling as long as 5,000 times, corresponding to a capacity fading of 0.0071% per cycle (Fig. 5d). Compared with the conventional $Na_{0.7}Fe_{0.7}Mn_{0.3}O_2$ nanoparticles, which were synthesized as a control sample, our $Na_{0.7}Fe_{0.7}Mn_{0.3}O_2$ mesoporous nanotubes exhibit much higher specific capacity and better cycling performance (Supplementary Figs 8 and 9). Another comparison with those previously reported results for sodium-ion batteries reveals that our $Na_{0.7}Fe_{0.7}Mn_{0.3}O_2$ mesoporous nanotubes demonstrate superior electrochemical performance (Supplementary Fig. 8m).

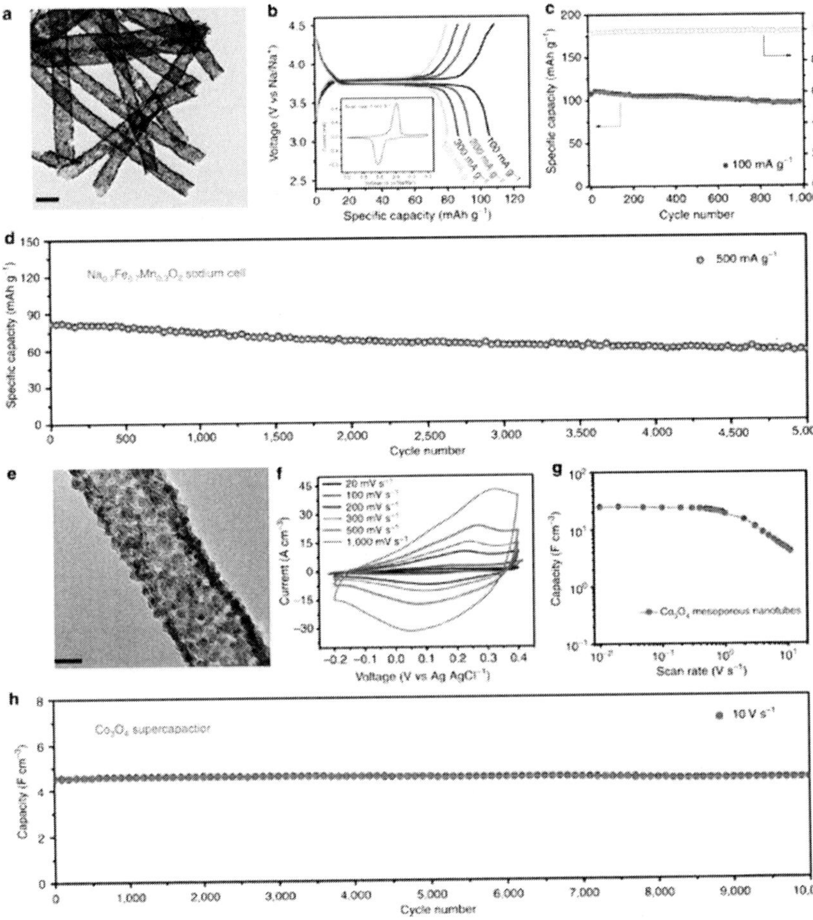

Figure 5. Characterization and electrochemical performance in sodium-ion batteries and supercapacitors. (**a**) TEM image of the $Na_{0.7}Fe_{0.7}Mn_{0.3}O_2$ mesoporous nanotubes with a scale at 200 nm. (**b**) Charge–discharge curves of $Na_{0.7}Fe_{0.7}Mn_{0.3}O_2$ measured at 100, 200, 300

and 500 mA g^{-1}, respectively. The inset is the CV collected at a scan rate of 5 mV s^{-1} in the potential range 3.0–4.5 V. (**c,d**) Cycling performance of $Na_{0.7}Fe_{0.7}Mn_{0.3}O_2$ mesoporous nanotubes tested for 1,000 cycles at 100 mA g^{-1} and for 5,000 cycles at 500 mA g^{-1} (**e**) TEM image of Co_3O_4 mesoporous nanotubes with scale bar at 20 nm. (**f**) CV curves obtained at different scan rates from 20, 100, 300, 500 to 1,000 mV s^{-1}, respectively. (**g**) Stack capacitance of Co_3O_4 mesoporous nanotubes versus scan rate. (**h**) Long cycling performance of Co_3O_4 mesoporous nanotubes tested for 10,000 times at a high rate of 10 V s^{-1}.

The Co_3O_4 mesoporous nanotubes are very uniform with a diameter of ~50 nm, and only contain ~5 nm nanoparticles on the tubes (Fig. 5e, Supplementary Fig. 10). A micro-supercapacitor device based on these nanotubes was fabricated on a silicon wafer (The fabrication process of micro-supercapacitor device is clearly illustrated in Supplementary Fig. 10). These cyclic voltammograms reveal the exceptionally enhanced electrochemical performance of this material (Fig. 5f). 25.0, 24.0 and 18.9 F cm^{-3} for stack capacitance are obtained at the scan rates of 0.01, 0.1 and 1 V s^{-1}, respectively (Fig. 5g). When tested at a high rate of 10 V s^{-1}, it delivers a capacity of 4.5 F cm^{-3}, and 98% of the initial capacity is maintained after cycling as long as 10,000 times (Fig. 5h). All these unveil that Co_3O_4 mesoporous nanotubes demonstrate excellent electrochemical performance with respect to both high rate and long life, showing great application potential.

3. DISCUSSION

In summary, our findings clearly indicate that the feasible gradient-electrospinning and controlled-pyrolysis methodology, with extensive material diversity, low cost, good repeatability and high yield, provides an efficient strategy to obtain controllable nanotubes for various inorganic materials, which can break through the limitation of the crystal growth orientation of each sample. On the basis of the formation mechanism, we successfully applied this technique to synthesize various multi-element oxides ($Li_3V_2(PO_4)_3$, $Na_3V_2(PO_4)_3$, $Na_{0.7}Fe_{0.7}Mn_{0.3}O_2$ and $LiNi_{1/3}Co_{1/3}Mn_{1/3}O_2$), binary-metal oxides ($LiMn_2O_4$, $LiCoO_2$, $NiCo_2O_4$ and LiV_3O_8) and single-metal oxides mesoporous nanotubes (CuO, Co_3O_4, SnO_2 and MnO_2). Therefore, we believe this technique could lead to rapid advancements in the development of 1D nanostructures. In addition, these novel mesoporous and pea-like nanotubes, which exhibit excellent electrochemical performance in lithium-ion batteries, sodium-ion batteries and supercapacitors, owing to their large surface area, high conductivity and robust structural stability, will have great potentials in not only the energy storage field, but also in numerous other frontiers.

4. METHODS

4.1 Electrospinning $Li_3V_2(PO_4)_3$ mesoporous nanotubes

First, the uniform precursor solution for electrospinning was prepared with low-molecular-weight (98–99% hydrolysed), middle-molecular-weight (86–89% hydrolysed) and high-molecular-weight (98–99% hydrolysed) PVA in a weight ratio of 3:2:1 in 20 ml of deionized water. (This same uniform precursor solution was used to prepare the following samples.) $LiOH{\cdot}H_2O$, NH_4VO_3 and $NH_4H_2PO_4$ were also added in a molar ratio of 3:2:3. After the mixture was stirred at 50 °C for 5 h, a viscous, uniform, transparent yellow precursor solution was obtained. The concentration of the precursor solution was ~9.5 wt%, the average viscosity was measured as ~0.53 dl g^{-1}. The relative humidity was maintained at a constant of ~40% RH, and the temperature was maintained at ~20 °C. The precursor solution was subsequently electrospun at a constant flow rate of~1 ml h^{-1} and at a high voltage of 20 kV (electrospinning equipment: SS-2534H from UCALERY Co., Beijing, China). The distance between the injector nozzle and the receiver was 15 cm. The composite nanowires were collected on revolving aluminium foil. After drying at 80 °C for 12 h, the composite nanowires were presintered at 300 °C (10 °C min^{-1}) in air for 3 h. The sample was then annealed at 800 °C (5 °C min^{-1}) under an argon atmosphere for 6 h to fully carbonize the decomposed PVA. Finally, uniform $Li_3V_2(PO_4)_3$ mesoporous nanotubes were obtained with a 7%carbon content (Supplementary Fig. 4).

4.2 Electrospinning $Na_3V_2(PO_4)_3$ mesoporous nanotubes

NH_4VO_3 and $NaH_2PO_4{\cdot}2H_2O$ were added in a molar ratio of 2:3 to the prepared PVA precursor. Then the precursor was electrospun at a high voltage of 20 kV. After the electrospinning process, the composite nanowires were sintered at 300 °C (10 °C min^{-1}) in air for 5 h and then annealed at 800 °C (5 °C min^{-1}) in an argon atmosphere for 5 h.

4.3 Electrospinning $Na_{0.7}Fe_{0.7}Mn_{0.3}O_2$ mesoporous nanotubes

$NaNO_3$, $Fe(NO_3)_3{\cdot}9H_2O$, and $Mn(CH_3COO)_2{\cdot}4H_2O$ were added in a molar ratio of 7:7:3 to the prepared PVA precursor. Then the precursor was electrospun at a high voltage of 20 kV. After the electrospinning process, the composite nanowires were first sintered at 300 °C (10 °C min^{-1}) in air for 5 h and then annealed at 700 °C (5 °C min^{-1}) under argon for 8 h.

4.4 Electrospinning $LiNi_{1/3}Co_{1/3}Mn_{1/3}O_2$ mesoporous nanotubes

$CH_3COOLi{\cdot}2H_2O$, $Ni(CH_3COO)_2{\cdot}4H_2O$, $Mn(CH_3COO)_2{\cdot}4H_2O$ and $Co(CH_3COO)_2{\cdot}4H_2O$ were added in a molar ratio of 3:1:1:1. Then the precursor was electrospun at a high voltage of 20 kV. After the electrospinning process, the composite nanowires were annealed at 700 °C (5 °C min^{-1}) in air for 5 h.

4.5 Electrospinning $LiMn_2O_4$ mesoporous nanotubes

$CH_3COOLi{\cdot}2H_2O$ and $Mn(CH_3COO)_2{\cdot}4H_2O$ were added in a molar ratio of 1:2 to the prepared PVA precursor. The precursor was electrospun at a high voltage of 18 kV. After the electrospinning process, the composite nanowires were annealed at 700 °C (5 °C min^{-1}) in air for 5 h.

4.6 Electrospinning $LiCoO_2$ mesoporous nanotubes

$CH_3COOLi{\cdot}2H_2O$ and $Co(CH_3COO)_2{\cdot}4H_2O$ were added in a molar ratio of 1:1 to the prepared PVA precursor. The precursor was electrospun at a high voltage of 18 kV. After the electrospinning process, the composite nanowires were sintered at 500 °C (10 °C min^{-1}) in air for 5 h.

4.7 Electrospinning $NiCo_2O_4$ mesoporous nanotubes

$Ni(CH_3COO)_2 \cdot 4H_2O$ and $Co(CH_3COO)_2 \cdot 4H_2O$ were added in a molar ratio of 1:2 to the prepared PVA precursor. The precursor was electrospun at a high voltage of 18 kV. After the electrospinning process, the composite nanowires were sintered at 400 °C (10 °C min^{-1}) in air for 5 h.

4.8 Electrospinning LiV_3O_8 mesoporous nanotubes

$LiOH \cdot H_2O$ and NH_4VO_3 were added in a molar ratio of 1:3. The precursor was electrospun at a high voltage of 18 kV. After the electrospinning process, the composite nanowires were sintered at 450 °C (10 °C min^{-1}) in air for 5 h.

4.9 Electrospinning CuO mesoporous nanotubes

A moderate amount of $Cu(NO_3)_2$ was added to the prepared PVA precursor. The precursor was electrospun at a high voltage of 16 kV. After the electrospinning process, the composite nanowires were sintered at 350 °C (10 °C min^{-1}) in air for 5 h.

4.10 Electrospinning Co_3O_4 mesoporous nanotubes

A moderate amount of $Co(CH_3COO)_2 \cdot 4H_2O$ was added to the prepared PVA precursor. The precursor was electrospun at a high voltage of 16 kV. After electrospinning, the composite nanowires were sintered at 350 °C (10 °C min^{-1}) in air for 5 h. The carbon from PVA was removed via the sintering process.

4.11 Electrospinning MnO_2 mesoporous nanotubes

A moderate amount of $Mn(CH_3COO)_2 \cdot 4H_2O$ was added to the prepared PVA precursor. Then the precursor was electrospun at a high voltage of 16 kV. After the electrospinning process, the composite nanowires were sintered at 350 °C (10 °C min^{-1}) in air for 5 h.

4.12 Electrospinning SnO_2 mesoporous nanotubes

A moderate amount of $SnCl_4$ was added to the prepared PVA precursor. The precursor was electrospun at a high voltage of 16 kV. After the electrospinning process, the composite nanowires were sintered at 500 °C (10 °C min^{-1}) in air for 5 h.

4.13 Electrospinning $Li_3V_2(PO_4)_3$ pea-like nanotubes

The preliminary electrospinning process was the same as that used to prepare the mesoporous nanotubes. After drying at 80 °C for 12 h, the composite nanowires were directly and immediately heated for 1.5 h in a furnace preheated to 300 °C. After the composite nanowires were annealed at 800 °C (5 °C min^{-1}) under an argon atmosphere for 6 h, uniform pea-like $Li_3V_2(PO_4)_3$ nanotubes with 19% carbon content were obtained (Supplementary Fig. 4).

4.14 Electrospinning Co pea-like nanotubes

An amount of $Co(CH_3COO)_2 \cdot 4H_2O$ was added to the prepared PVA precursor. Then the precursor was electrospun at a high voltage of 16 kV. After the electrospinning process, the composite nanowires were presintered in a muffle furnace at 300 °C for 1.5 h and then annealed at 600 °C under argon for 5 h.

4.15 Electrospinning $LiCoO_2$ pea-like nanotubes

$CH_3COOLi \cdot 2H_2O$ and $Co(CH_3COO)_2 \cdot 4H_2O$ were added in a molar ratio of 1:1 to the prepared PVA precursor. Then the precursor was electrospun at a high voltage of 18 kV. After the electrospinning process, the composite nanowires were presintered in a muffle furnace at 300 °C for 1.5 h then annealed at 600 °C (5 °C min^{-1}) under argon for 5 h.

4.16 Electrospinning $Na_{0.7}Fe_{0.7}Mn_{0.3}O_2$ pea-like nanotubes

$NaNO_3$, $Fe(NO_3)_3 \cdot 9H_2O$, and $Mn(CH_3COO)_2 \cdot 4H_2O$ were added in a molar ratio of 7:7:3. Then the precursor was electrospun at a high voltage of 20 kV. After the electrospinning process, the composite nanowires were presintered in a muffle furnace at 300 °C for 1.5 h and then annealed at 700 °C (5 °C min^{-1}) under argon for 8 h.

4.17 Morphology and structure characterizations

The crystallographic information of the final products was measured using a Bruker D8 Advance X-ray diffractometer equipped with a Cu Kα radiation source; the samples were scanned over the 2θ range from 10° to 80° at room temperature. SEM images were collected using a JEOL-7100F SEM, and TEM images were collected using a JEM-2100F TEM. The BET surface area was calculated from nitrogen adsorption isotherms measured at 77 K using a Tristar-3020 instrument. Energy-dispersive X-ray spectra were recorded using an Oxford IE250 system. thermogravimetric–DSC analyses were conducted using a STA-449C. X-ray photoelectron spectroscopy analysis was conducted on a VG Multilab 2000. Raman spectra were obtained using a Renishaw INVIA micro-Raman spectroscopy system. The surface tension was tested by an automatic surface tensiometer (CC2L202) from 2 to 200 N m^{-1}.

4.18 Preparation of $Li_3V_2(PO_4)_3$ lithium half cells

The 2016 coin cells were assembled in a glovebox filled with pure argon gas. Lithium foil was used as the anode, and a solution of $LiPF_6$ (1 M) in

EC/DEC (1:1 vol/vol) was used as the electrolyte. The cathode was composed of a ground mixture of 70% $Li_3V_2(PO_4)_3$ active material, 20% acetylene black and 10% poly(tetrafluoroethylene). After coating onto aluminium foil, the cathode was cut into round slice with ~0.36 cm^2 in area and ~0.1 mm in thickness. And the loading rate of $Li_3V_2(PO_4)_3$ active material was ~2 mg for one 2016 coin cell.

4.19 Preparation of $Li_3V_2(PO_4)_3/Li_4Ti_5O_{12}$ full batteries

The $Li_3V_2(PO_4)_3/Li_4Ti_5O_{12}$ battery was also assembled in a glovebox filled with pure argon gas. $Li_3V_2(PO_4)_3$ was coated onto aluminium foil as the cathode, and $Li_4Ti_5O_{12}$ was coated onto copper foil as the anode.

4.20 Preparation of $Na_{0.7}Fe_{0.7}Mn_{0.3}O_2$ sodium batteries

The 2016 coin cells were assembled in a glovebox filled with pure argon gas. Sodium foil was used as the anode, and 1 M $NaClO_4$ in a mixture of ethylene carbonate/dimethyl carbonate (1:2 in weight) with trace propylene carbonate (electrolyte additive) was used as the electrolyte. The prepared cathodes contained 70% $Na_{0.7}Fe_{0.7}Mn_{0.3}O_2$ active material, 20% acetylene black and 10% polyvinylidene difluoride (using *N*-methylpyrrolidone as the solvent). After coating onto aluminium foil, the cathode was cut into round slice with ~0.36 cm^2 in area and ~0.1 mm in thickness, and the loading rate of $Na_{0.7}Fe_{0.7}Mn_{0.3}O_2$ was ~2.5 mg for one 2016 coin cell.

4.21 Preparation of Co_3O_4 supercapacitors

Electrochemical measurements were performed within the potential window of −0.2 to 0.4 V using an Autolab potentiostat/galvanostat (Autolab PGSTAT 302 N) in a three-electrode configuration with 1 M NaOH aqueous solution as the electrolyte; the measurements were performed at room temperature. A 1-cm^2 piece of silicon wafer on which the Co_3O_4 mesoporous nanotubes were deposited was directly used as the working electrode (Supplementary Fig. 10). The reference and counter electrodes were an Ag/KCl electrode and a platinum plate, respectively.

4.22 Electrochemical measurements

Galvanostatic charge/discharge measurements were performed using a multichannel battery testing system (LAND CT2001A). Cyclic voltammograms and electrochemical impedance spectra were collected at room temperature using an Autolab potentiostat/galvanostat.

REFERENCES

1. Armand, M. & Tarascon, J.-M. Building better batteries. *Nature* **451**, 652–657 (2008).

2. Noorden, R. V. A better battery. *Nature* **507**, 26–28 (2014).

3. Wu, H. *et al.* Stable cycling of double-walled silicon nanotube battery anodes through solid–electrolyte interphase control. *Nat. Nanotechnol.* **7**, 310–315 (2012).

4. Zhao, Y. & Jiang, L. Hollow micro/nanomaterials with multilevel interior structures. *Adv. Mater.* **21**, 3621–3638 (2009).

5. Huang, Y., Duan, X. F., Wei, Q. Q. & Lieber, C. M. Directed assembly of one-dimensional nanostructures into functional networks. *Science* **291**, 630–633 (2001).

6. Jung, H. G., Jang, M. W., Hassoun, J., Sun, Y. K. & Scrosati, B. A high-rate long-life $Li_4Ti_5O_{12}/Li[Ni_{0.45}Co_{0.1}Mn_{1.45}]O_4$ lithium-ion battery. *Nat. Commun.* **2**, 516 (2011).

7. Yang, F. *et al.* Chirality-specific growth of single-walled carbon nanotubes on solid alloy catalysts. *Nature* **510**, 522–524 (2014).

8. Qin, Y., Lee, S.-M., Pan, A. L., Gösele, U. & Knez, M. Rayleigh-instability-induced metal nanoparticle chains encapsulated in nanotubes produced by atomic layer deposition. *Nano Lett.* **8**, 114–118 (2008).

9. Lee, S. W. *et al.* High-power lithium batteries from functionalized carbon-nanotube electrodes. *Nat. Nanotechnol.* **5**, 531–537 (2010).

10. Wang, X., Zhuang, J., Peng, Q. & Li, Y. D. A general strategy for nanocrystal synthesis.*Nature* **43**, 121–124 (2005).

11. Magasinski, A. *et al.* High-performance lithium-ion anodes using a hierarchical bottom-up approach. *Nat. Mater.* **9**, 353–358 (2010).

12. Deng, Y. H., Wei, J., Sun, Z. K. & Zhao, D. Y. Large-pore ordered mesoporous materials templated from non-Pluronic amphiphilic block copolymers. *Chem. Soc. Rev.* **42**,4054–4070 (2013).

13. Yao, J., Yan, H. & Lieber, C. M. A nanoscale combing technique for the large-scale assembly of highly aligned nanowires. *Nat. Nanotechnol.* **8**, 329–335 (2013).

14. Chen, C. *et al.* Highly crystalline multimetallic nanoframes with three-dimensional electrocatalytic surfaces. *Science* **343**, 1339–1343 (2014).

15. Joshi, R. K. & Schneider, J. J. Assembly of one dimensional inorganic nanostructures into functional 2D and 3D architectures. Synthesis, arrangement and functionality. *Chem. Soc. Rev.* **41**, 5285–5312 (2012).

16. Yaman, M. *et al.* Arrays of indefinitely long uniform nanowires and nanotubes. *Nat. Mater.***10**, 494–501 (2011).

17. Ren, Y., Ma, Z. & Bruce., P. G. Ordered mesoporous metal oxides: synthesis and applications. *Chem. Soc. Rev.* **41**, 4909–4927 (2012).

18. Xia, Y. N. *et al.* One-dimensional nanostructures: synthesis, characterization and applications. *Adv. Mater.* **15**, 353–389 (2003).

19. Li, D. & Xia, Y. N. Electrospinning of nanofibers: reinventing the wheel? *Adv. Mater.* **16**,1151–1170 (2004).

20. Li, D., MaCann, J. T. & Xia, Y. N. Electrospinning: a simple and versatile technique for producing ceramic nanofibers and nanotubes. *J. Am. Ceram. Soc.* **89**, 1861–1869 (2006).

21. Yuan, C. Z., Wu, H. B., Xie, Y. & Lou, X. W. Mixed transition-metal oxides: design, synthesis, and energy-related applications. *Angew. Chem. Int. Ed.* **53**, 1488–1504 (2014).

22. Mai, L. Q. *et al.* Electrospun ultralong hierarchical vanadium oxide nanowires with high performance for lithium ion batteries. *Nano Lett.* **10**, 4750–4755 (2010).

23. Wu, J., Wang, N., Zhao, Y. & Jiang, L. Electrospinning of multilevel structured functional micro-/nanofibers and their applications. *J. Mater. Chem. A* **1**, 7290–7305 (2013).

24. Zussman, E., Yarin, A. L., Bazilevsky, A. V., Avrahami, R. & Feldman, M. Electrospun polyacrylonitrile/poly(methyl methacrylate)-derived turbostratic carbon micro-/nanotubes.*Adv. Mater.* **18**, 348–353 (2006).

25. Zhan, S. H., Chen, D. R., Jiao, X. L. & Tao, C. H. Long TiO_2 hollow fibers with mesoporous walls: sol-gel combined electrospun fabrication and photocatalytic properties. *J. Phys. Chem. B* **110**, 11199–11204 (2006).

26. Hou, H. L. *et al.* General strategy for fabricating thoroughly mesoporous nanofibers. *J. Am. Chem. Soc.* **136**, 16716–16719 (2014).

27. Sun, Y. G., Mayers, B. & Xia, Y. N. Metal nanostructures with hollow interiors. *Adv. Mater.***15**, 641–646 (2003).

28. Li, D. & Xia, Y. N. Direct fabrication of composite and ceramic hollow nanofibers by electrospinning. *Nano Lett.* **4**, 933–938 (2004).

29. McCann, J. T., Marquez, M. & Xia, Y. N. Melt coaxial electrospinning: a versatile method for the encapsulation of solid materials and fabrication of phase change nanofibers. *Nano Lett.* **6**, 2868–2872 (2006).

30. Zhao, Y., Cao, X. Y. & Jiang, L. Bio-mimic multichannel microtubes by a facile method. *J. Am. Chem. Soc.* **129**, 764–765 (2007).

31. Shin, Y. M., Hohman, M. M., Brenner, M. P. & Rutledge, G. C. Experimental characterization of electrospinning: the electrically forced jet and instabilities. *Polymer (Guildf).* **42**, 9955–9967 (2001).

32. Fridrikh, S. V., Yu, J. H., Brenner, M. P. & Rutledge, G. C. Controlling the fiber diameter during electrospinning. *Phys. Rev. Lett.* **90**, 144502 (2003).

33. Wu, Z. X. *et al.* A general "surface-locking" approach toward fast assembly and processing of large-sized, ordered, mesoporous carbon microspheres. *Angew. Chem. Int. Ed.* **52**,13764–13768 (2013).

34. Theron, A., Zussman, E. & Yarin, A. L. Electrostatic field-assisted alignment of electrospun nanofibres. *Nanotechonolgy* **12**, 384–390 (2001).

35. Liu, H. *et al.* Capturing metastable structures during high-rate cycling of $LiFePO_4$nanoparticle electrodes. *Science* **344**, 1480 (2014).

36. Yang, P. D., Zhao, D. Y., Margolese, D. I., Chmelka, B. F. & Stucky, G. D. Generalized syntheses of large-pore mesoporous metal oxides with semicrystalline frameworks. *Nature***396**, 152–155 (1998).

37. Kang, B. & Ceder, G. Battery materials for ultrafast charging and discharging. *Nature* **458**,190–193 (2009).

38. Nishijima, M. *et al.* Accelerated discovery of cathode materials with prolonged cycle life for lithium-ion battery. *Nat. Commun.* **5**, 4553 (2015).

39. Mai, L. Q., Tian, X. C., Xu, X., Chang, L. & Xu, L. Nanowire electrodes for electrochemical energy storage devices. *Chem. Rev.* **114**, 11828–11862 (2014).

40. Masquelier, C. & Croguennec, L. Polyanionic (phosphates, silicates, sulfates) frameworks as electrode materials for rechargeable Li (or Na) batteries. *Chem. Rev.* **113**, 6552–6591(2013).

41. Yabuuchi, N. *et al.* P2-type $Na_x[Fe_{1/2}Mn_{1/2}]O_2$ made from earth-abundant elements for rechargeable Na batteries. *Nat. Mater.* **11**, 512–517 (2012).

42. Kalluri, S. *et al.* Electrospun P2-type $Na_{2/3}(Fe_{1/2}Mn_{1/2})O_2$ hierarchical nanofibers as cathode material for sodium-ion batteries. *ACS Appl. Mater. Interfaces* **6**, 8953–8958(2014).

43. Xu, J. *et al.* Identifying the critical role of Li substitution in P2-$Na_x[Li_yNi_zMn_{1-y-z}]O_2$ $(0<x, y, z<1)$ intercalation cathode materials for high-energy Na-ion batteries. *Chem. Mater.* **26**,1260–1269 (2014).

44. Oh, S. M. *et al.* Advanced $Na[Ni_{0.25}Fe_{0.5}Mn_{0.25}]O_2/C\text{-}Fe_3O_4$ sodium-ion batteries using EMS electrolyte for energy storage. *Nano Lett.* **14**, 1620–1626 (2014).

45. Wu, Z. S., Parves, K., Feng, X. L. & Müllen, K. Graphene-based in-plane micro-supercapacitors with high power and energy densities. *Nat. Commun.* **4**, 2487 (2013).

9

MICROWAVE DIELECTRIC PROPERTIES OF CARBON NANOTUBE COMPOSITES

L. Liu[1], L. B. Kong[1], W. Y. Yin[2], Y. Chen[3] and S. Matitsine[1]

[1] Temasek laboratories, National University of Singapore,

[2] Centre for Optical and EM Research, State Key Lab of MOI, Zhe Jiang University Hangzhou,

[3] School of Chemical and Biomedical Engineering, Nanyang Technological University,

1. INTRODUCTION

Carbon black (CB) (Donnet et al., 1993) and carbon fiber (CF) (Burchell, 1999) have been used as fillers in radio wave and microwave composites for more than half century. Typical applications include electromagnetic compatibility (EMC) or electromagnetic interference (EMI), microwave absorbing coatings, and anti-electrostatic materials, *etc* due to their good electric/dielectric properties, stability and chemical resistance as compared to composites with metal powders of nano or micron meter in size. CB composites normally have frequency-dependent permittivity which is useful in the design of broadband absorbing or functional materials at microwave band. The frequency dispersion in CB composites is mainly caused by the effect of large agglomerate of CB spherical particles with diameter below 100 nanometers and usually appears in composites with large concentration of fillers. CF composites can have special microwave properties, such as high permittivity at relatively low concentrations of fibers (Lagarkov et al, 1998). The dispersive property of CF composites at microwave band is determined by the frequency-dependent response of individual fibers or fiber clusters (Liu, et al, 2007a). The dispersion can be observed even at very low concentration of fibers, for example, less than one percent in volume. This feature makes CF composites advantageous for many technical applications, such as lightweight absorbing materials (Neo & Varadan, 2004) and phantom materials for modeling the electromagnetic response of biological issue (Youngs, et al, 2002). However, to obtain strong dielectric dispersion at microwave frequencies, CF must contain fibers which are at least millimeters in length to achieve large aspect ratio considering the diameter of commercial CF which is normally close to 10 microns. The large dimension of the fibrous inclusions restricts the practical application of CF composites. Also, the favorite processing methods, such as spray technique, are not applicable for CF composites. The planar CF composites are anisotropic with degraded dielectric response across the plane. Due to the heterogeneity of the composites, a surface layer of millimeters in thickness appears with properties different from those of the bulk composites (Simovski, et al, 2000 andMatitsine, et al, 2003). This gives rise to difficulty in modeling and designing composites with desired properties.

Carbon nanotubes (CNTs) have been fabricated since the early 1990s (Ijima, 1991). The primary research interests include the synthesis or growth of CNTs, because it was a challenging task to prepare enough amounts of CNTs with desired dimension and purity for measurement purpose during the early stage. With the development of processing method like arc discharge and laser ablation of graphite, as well as more productive chemical vapor deposition (CVD) and plasma enhanced CVD method, high purity CNTs with controllable wall-thickness/length and acceptable price are commercially available now (Meyyappan, 2005). Other research interests include the characterization and theoretical understanding of their basic properties of CNTs. The physical properties of carbon nanotubes were reviewed in literature (Saito, 1998), which have been considered as the classic monograph on CNTs. Due to the unique electronic structure depending on the diameter and chirality, CNTs could be the smallest semiconductor devices likely to be fabricated (Saito, 1998). And the extraordinary mechanical properties of CNTs make them suitable replacement of CF in continuous fiber reinforced composites for defence or aerospace applications. Other potential applications includes single- or multi-walled CNTs sensor for scanning probe microscopy, and biosensor or device applications such as CNT-based diodes, transistors and field emitter, *etc* (Meyyappan, 2005).

Due to their fibrous shape with extremely large in aspect ratio, CNTs may allow low resistivity and high permittivity and frequency dispersion to be obtained in composites with extremely low concentrations. As compared with CF, one significant difference is that CNTs are nanoscaled particles whose diameter is several orders of magnitude smaller than that of CF (typically about 10 μm). It is well known that the nanosized particles usually display distinct properties from microsized particles of the same composition, which is the primary reason for the great attention currently given to the radio and microwave frequency performance of CNTs composites. A number of novel features have been reported on CNTs in the literature, these results demonstrate the possibility to design CNTs composites with electric/dielectric properties which are more diverse than those obtainable with other carbon fillers. That is why in the last few years, main research interest in CNT was shifted from fabrication methods to the novel properties, such as the conductivity at low concentration and dielectric properties at low frequency (DC to 1MHz) and high frequency (above 1MHz, including radio wave, microwave, millimeter wave, Terahertz and Inferred wave frequency). Since the dielectric properties of CNT composites at microwave frequency are the main focus of this chapter, previous research works on the electric properties, dielectric permittivity at high frequency, as well as potential EMC/EMI and absorbing applications which are most relevant to our research objective, will be reviewed sequentially.

After the observation of a conductivity threshold in polymer/CNTs composites (Coleman, et al, 1998), around 200 publications have reported on the electrical percolation threshold of CNTs in different polymer systems by 2009 (W. Bauhofer & J. Z. Kovacs, 2009). Different CNTs, synthesis methods, dispersion treatment and polymer matrix have been employed in the previous research. Since it is established that the percolation theory or power law could explain the electric conductivity of CNTs composite (Martin et al, 2004). Minimum percolation ratio and maximum conductivity at low frequency becomes the main research focus. Due to the dispersion method and difference in the holding matrix, percolation varies from less than 0.01 percent to a few percent. Theoretically, typical aspect ratio of 1000 produces percolation value about 0.1% (Balberg et al, 1984). Practically, a few order of difference could be found from composite with same CNTs and different polymer matrix (W. Bauhofer & J. Z. Kovacs, 2009). Maximum conductivity of MWCNTs composite could be realized when the weight concentration is larger than 10% (Skakalove et al, 2005). And the maximum conductivity seems insensitive to the types of CNTs and treatments. Nevertheless, it is a challenging task to disperse CNTs uniformly at such high concentration. The mechanism of charge transport in CNTs composites could be explained by hopping and tunneling effect among neighboring tubes (Kim, 2006). However, the morphology of CNTs in holding matrix seems difficult to be represented by the classical percolation theory.

Permittivity (ɜ=ɜ'-*j* ɜ'''') instead of AC conductivity is normally used to describe the dielectric properties of CNTs composites as frequency is above 1MHz. Also, measurement methodologies and potential applications could be different from that at low frequency. Microwave properties are normally measured with impedance or network analyzer through coaxial line or waveguide fixture. A number of special features have been reported on CNT in the literature, such as excitation of localized electronic states resulting in high intrinsic permittivity (Watts et al, 2003).The microwave permittivity is found to exhibit smooth frequency-dependent (Browning, et al, 1998& Sandler, et al, 2003). These results seem to demonstrate the possibility to obtain CNT composites with microwave properties which are more diverse than those obtainable with other types of carbon fillers, e. g., with various types of dispersive dielectrics, with high dielectric but having relative low dielectric loss (Yang, et al, 2009), and tunable dielectric under small bias voltage (Liu, et al, 2008). In addition, the nanoscaled size of CNT in composites may avoid the drawbacks of CF composites, which are limited by the macroscopic size of the fibers for thin structure and component applications.

Since the low resistivity of CNTs composite with low filling factor as compared with CB composite which are commonly employed as filler of rubber, conductivity polymer and shielding coating, the shielding effectiveness (SE) of CNTs composites are mostly investigated at microwave frequency (Huang et al, 2007). MWCNTs/Ceramic composites were measured at Ku-band with SE up to 25 to 30dB achieved (Shi & Liang, 2008). Some experimental results showed that absorption is the major shielding mechanism and reflection is the secondary mechanism (M. H. Al-saleh & U. Sundararaj, 2009). SWCNTs thin film demonstrated good shielding of EM wave at terahertz frequency while still maintaining good transparency for visible light (Seo, et al, 2008). Measurement results show that CNT sheets are more effective in providing EMI shielding compared with graphite and carbon black sheets over broad frequency range due to better electron transmission (Wang, et al, 2009). Radar absorbing structures with MWCNTs and polyurethane foam were designed with optimization method before measured with free space setup (Park, et al, 2006). 10 dB reflection loss at X-band could be achieved by MWCNTs layer of thickness about 3.3mm (Lee et al, 2006). Self-sensing and self-actuation response were found out from CNTs composites which could be explained by electrostriction effect (Lee & Shkel, 2007). The tunability of both real and imaginary parts of the permittivity found from CNTs composites has potential applications for smart materials and structures (Liu, et al, 2008).

The shielding effectiveness and absorption of EM wave are functions of electric and dielectric properties; the objective of this chapter is to investigate the resistivity and dielectric permittivity of CNTs composites for different wall thicknesses. The organization of this chapter is as follow. Theoretical methods and models concerning dielectric properties of CNTs composites are reviewed briefly in Section 2. The materials, fabrication conditions and measurement method of dielectric properties are addressed in Section 3. The results in Section 4 include electric properties, frequency-dependent permittivity, tunable permittivity under bias voltage and potential applications. Last but not least, the conclusion of this research work are summarized in Section 5.

2. THEORY

In CB composites, the microwave behavior is attributed mainly to the large agglomerates of carbon particles. In this case, the properties of the composite are consistent with the percolation theory that accounts for the statistical properties of clusters comprising many inclusions in electrical contact within the composite (Bergman & Imry, 1977). The percolation approach (Satuffer & Aharony, 1991) depends on the empirical percolation threshold *p c* that cannot be derived from the percolation theory. The percolation threshold is defined as the least concentration at which a composite with conductive inclusions is capable of conducting direct current. For the permittivity (=+i) of a composite comprising conductive inclusions and a lossless dielectric host matrix of permittivity $_{h}$, percolation theory predicts a power dependence on the frequency f (Gefen, et al, 1983):

$$\varepsilon'' = A f^{-x}, \quad \varepsilon' - \varepsilon_{h} = B f^{-y}, \tag{1}$$

In (1), A and B are proportionality factors, x and y are critical exponents. In the classical percolation theory, the range of concentrations p where scaling law (1) is not expected to be observed in a wide frequency range, but typically very narrow, for example $(p-p_c)/p_c$ 0.1 (Gefen, et al, 1983). However, in many measured data of CB composites, the scaling frequency dispersion is observed in a wide range of both concentrations and frequencies (Brosseau, 2002). If (1) is valid over the entire frequency range, it follows from the Kramers–Kronig relations that (Jonscher, 1999):

$$x = y_{,} \quad \tan\delta = \frac{A}{B} = \tan\frac{\pi y}{2} \tag{2}$$

where tan is the dielectric loss tangent, and the second of (2) is valid for ' $>>_h$. Therefore, the critical exponents for ' and " are the same, while the dielectric loss tangent is frequency-independent. Equation (1) coincides with the "universal relaxation law" established for many types of dielectrics (Jonscher, 1999). In the percolation theory, the critical exponents are considered to be universal, i.e. the same for different composites, independent of the type and properties of inclusions comprising the composite. The physical basis is that the behavior of the composite is determined by the statistical properties of the clusters, but not by the parameters of individual inclusions.

The universal values for critical exponents of frequency dependence are usually derived from the exponents for concentration dependence. For this, two different approaches are known, with exponent value of either 0.27 or 0.42 (Song, et al, 1986). Many numerical and experimental data are obtained within these limits. For example, numerical studies of random RC networks (Straley, 1977) produce 0.28 for 3-D bond problem, but 0.18 for 3-D site problem. For continuum percolation problem, the value of 0.38 is obtained from the studies of water-oil emulsions (Bordi, et al, 1996). Therefore, the critical exponent values are believed to be within the range of 0.2 to 0.4, with the uncertainty attributed to the difference in geometry between the bond, site, and continuum percolation problems that may result in different properties of the clusters. For CNT composites, the critical exponent of 0.34, consistent with percolation theory, is obtained (Kim, et al, 2003). In all cases, x is less than 1. Therefore, the AC conductivity of the composite must increase with frequency.

In actual composites, the percolation dispersion governed by (1) appears within a limited range of frequency. The high frequency limit is determined by the frequency at which the conductivity of the composite is equal to the conductivity of the inclusions, and therefore, no longer increases with frequency. The low frequency limit is as considered

(Laibowitz & Gefen, 1984). Below this low frequency limit, the real permittivity or conductivity is independent of frequency at concentrations below p_c or above p_c correspondingly. Due to the finite frequency range for percolation dispersion, (2) is not exact for practical materials, with the discrepancy increases as the frequency approaches the limits of the percolation frequency range. If (1) is valid over restricted frequency range, then the dielectric loss tangent increases relatively with respect to the value given by (2), as the high frequency limit of this frequency range is approached. Near the low frequency limit, the loss tangent decreases for concentrations below p_c and increases for concentrations above p_c.

3. EXPERIMENT

3.1. Characterization Of Cnts

Single-, double- and multi-walled carbon nanotubes (SWCNT, DWCNT, and MWCNT) produced by CVD method were obtained from the Timesnano, Chengdu Organic Chemicals Co. Ltd., the Chinese Academy of Sciences. As produced CNTs were purified by the supplier with Oxidation and HCL wash method. The specifications of these CNTs are given in Table 1. Figure 1 shows the image of these CNTs with high resolution field emission SEM (JEOL JSM-6340F). The diameter of SWCNT and DWCNT seems larger than the values given by the supplier. Possible reasons can be the appearance of bundles or small amount of the CNTs may be multi-walled instead of single-walled. The aspect ratio is calculated based on the length over the average diameter of individual CNTs. It can also been observed from Figure 1 that length of CNTs are larger than the scope observed by SEM which is about a few micron meters.

The results of thermal gravimetric analysis (TGA) shown in Figure 2 verify that the carbon content in samples is within the ranges provided by the supplier. The residual weights (which is usually contributed by metallic catalyst particles) of the SWCNT and MWCNT samples are almost zero. It suggests that metallic particles can be efficiently removed by the HCl wash. However, about 6 wt% of residues is still remained on the DWCNT sample, shown in Figure 2(b). This may be caused by the different purification method applied by the supplier. Differential Thermal Gravimetry (DTG) results inFigure 2 also demonstrate that most of the carbon materials can be oxidized in a narrow temperature range centered at the peak oxidation temperature (Tg). The narrow oxidation temperature range suggests that carbon species in the samples have similar structures. T_g of the SWCNT sample is at 624 C, which is higher than the oxidation temperature of purified SWCNTs with diameter of 0.7-1.0 nm (Wei, et al, 2008). The increase of T_g may be contributed by the large diameter tubes or certain amount of impurities, such as MWCNTs. The decrease of T_g for DWCNT samples is caused by the metallic residues remained on the sample which may catalyze the oxidation of carbon and result in the lower oxidation temperature at 604 C. The DTG result of MWCNT shows a relative broader oxidation peak, as well as a higher T_g. The relative large range of wall thickness might contribute to the observed difference.

Table 1. Specifications of three types of CNTs.

	SWCNT	DWCNT	MWCNT
Purity	95%	90%	95%
Impurities	MWCNT, Ash, Metal	MWCNT, Ash, Metal	Ash, Metal
Length	~30 μm	~50 μm	~50 μm
Diameters	1~2nm	2~4nm	8~15nm
Aspect ratio	~20k	~17k	~2k

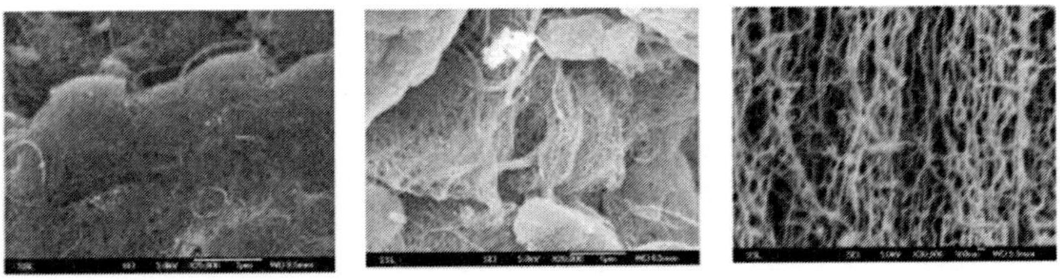

Figure 1. High-resolution SEM of CNT (a) SWCNT (b) DWCNT (c) MWCNT.

Raman spectroscopy was further applied to analyze the three CNT samples. Samples show three typical bands: radial breathing mode (RBM) between 150 and 350 cm^{-1} characteristic

for SWCNT or DWCNT, the D band at approximately 1300 cm^{-1} assigned to defective and disordered carbon species, and the peak complex centered around 1600 cm^{-1}, known as the G band, which is characteristic for ordered carbon species such as CNTs and graphite. The sharp RBM peaks from the SWCNT sample in Figure 3(a) and the DWCNT sample in Figure 3(b) verify the high content of SWCNTs and DWCNTs. The very low D band intensities also suggest that little amorphous carbon exists. Figure 3(c) shows the typical Raman spectrum of MWCNTs. However, it is difficult to quantify the purity of different CNT species. Instead, it is confirmed that there are significant amount of CNTs in the three samples.

3.2. Fabrication Of Cnts Composites

The host polymer to prepare composites was commercial two-component silicone (KE1800T) from Shin Etsu. CNTs and silicone (Component A) were dispersed in Toluene (Methylbenzene) with Ultrasonication processer (Sonics VCX500) for 20 min. The resistance of liquid mixture with 0.01g of SWCNTs, 1g of silicone and 25ml of Toluene was monitored with a parallel-plate sensor connected with insulation multimeter (Megger BMM80). The dependence of resistance on the sonication time plotted in Figure 4 shows that 20 minutes sonication time is probably enough to disperse the mixture to stable stage. A SEM image (shown in Figure 5) of SWCNTs after sonication process shows that big agglomerates in Figure 1(a) were broken and no big bundles (larger than 1μm) could be found.

Mixture of CNTs, silicone and Toluene in a 25ml crucible was heated up with a hotplate at 90 C for 8 hours to remove the solvent. After that, the 10% of liquid hardener (Component B) was added and mixed manually before pouring into mould. The mould was put into convection oven to cure at 70 C for 24 hours. As a control, another group of SWCNT, DWCNT and MWCNT was mixed manually with silicone component A and component B for 15 min before cured under the same conditions. And the control group prepared by manual mixing will be used as the benchmark of the sonication process.

Two types of samples were fabricated with moulding method. Disk shape samples have diameter of 16mm and thickness of 2mm to measure the resistivity and permittivity from 0.001GHz to 1GHz. Toroidal shape samples for permittivity measurement at high frequency (from 0.05GHz to 5GHz) have outer diameter of 14mm, inner diameter of 7mm and thickness of about 2mm.

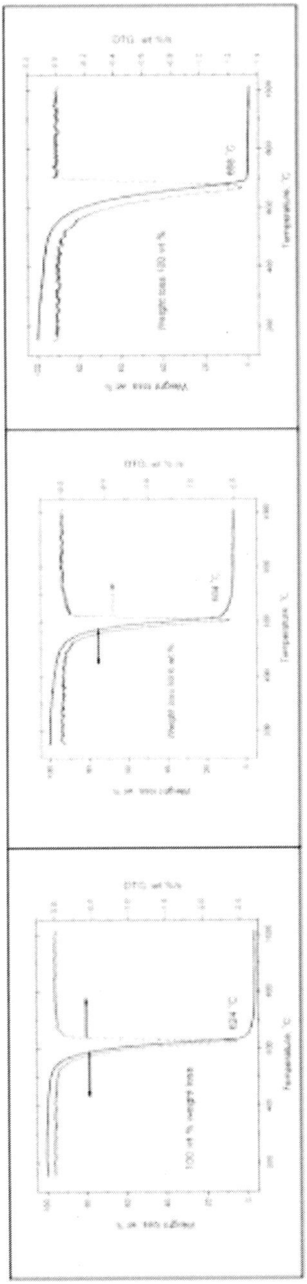

Figure 2. Purity analysis of (a) SWCNT (b) DWCNT (c) MWCNT with TGA.

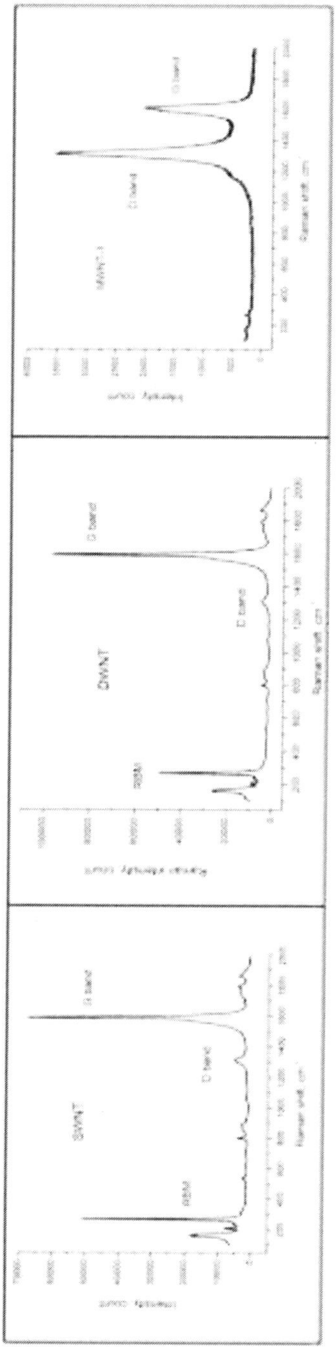

Figure 3. Raman spectrum of (a) SWCNT (b) DWCNT (c) MWCNT.

3.3. Measurement Methods

DC resistance of the samples was measured using a Megger™ BMM80™ insulation multimeter and a parallel plate fixture. Before measurement, the disk-shape samples were cleaned with ethanol to remove the release agent sprayed on the mould prior to casting. This is important since the agent contributed to the measured resistance. Resistivity was calculated from the measured resistance and dimension of the samples. Agilent 4991A RF impedance/materials analyzer was employed for permittivity measurement from 0.001GHz to 1 GHz. Figure 6 shows the test fixture (Agilent 16453A) for measurement of dielectric permittivity. During measurement, a bias voltage up to 40 volts was applied to the sample in between the upper and lower electrodes. The permittivity measured is in the direction of the bias voltage applied. Special care was taken to keep good electrical contact between the electrodes and the sample by applying certain pressure. Since there is no air gap effect, the impedance method can achieve good accuracy for measurement of conductive samples without correction [Liu, et al, 2007b, 2008].

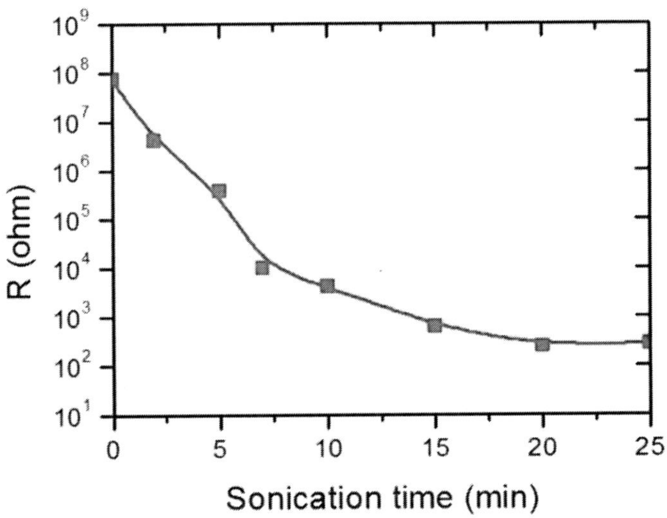

Figure 4. Resistance vs. sonication time.

To validate the measurement results by impedance analyzer, 14mm coaxial line fixture (as shown inFigure 7) and network analyzer (Agilent N5230A) was used to measure the permittivity of SWCNT composites samples from 0.05 to 5GHz. The permittivity was extracted from the measured S-parameters of the sample placed within the coaxial line. Since the outer diameter of the sample can be slightly smaller than the inner diameter of the test fixture due to the shrinkage of the sample after curing, the air gap in between is the largest contributor of uncertainty, especially when the concentration is high. Therefore, only 4wt% and below measured by the two different methods have been compared. The real and imaginary permittivity of SWCNTs composites of weight concentration from 0.5% to 4% are plotted in Figure 8. It shows rather good agreement between these two techniques over 0.05GHz to 1GHz for concentration up to 3%. Both real and imaginary permittivity measured by 14mm coaxial line fixture at 4% show lower values than these from impedance method, which could be properly explained through air gap effect (Liu et al, 2007b and reference therein). After correction of air gap effect, the real permittivity (dash line) shows better agreement with impedance measurement. It is proved that impedance measurement has better accuracy even at high concentration as compared with coaxial line method.

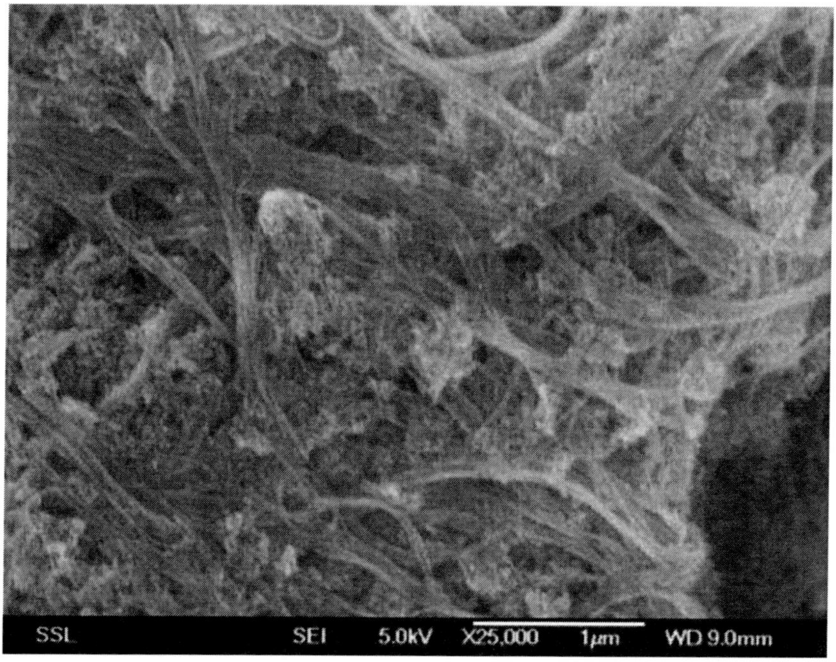

Figure 5. SWCNT after sonication.

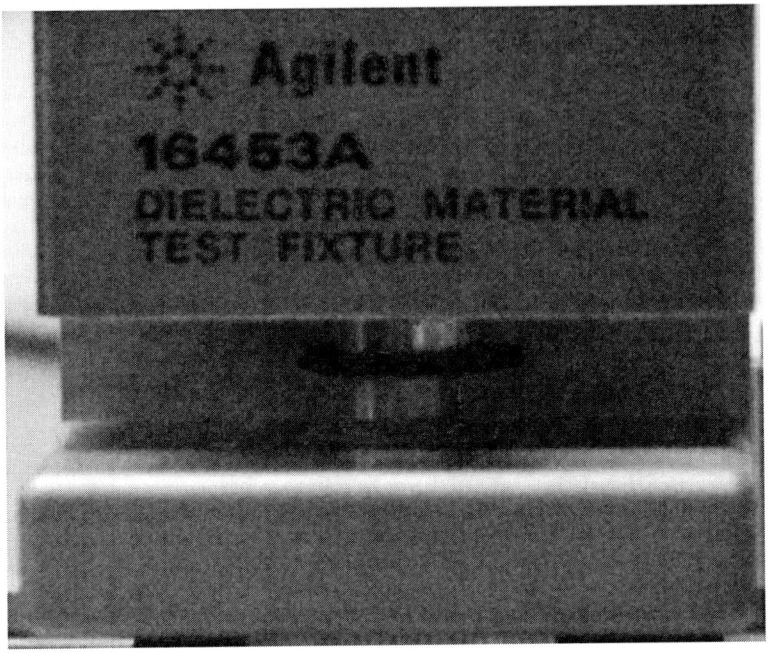

Figure 6. Test fixture and sample for permittivity from 1MHz to 1GHz.

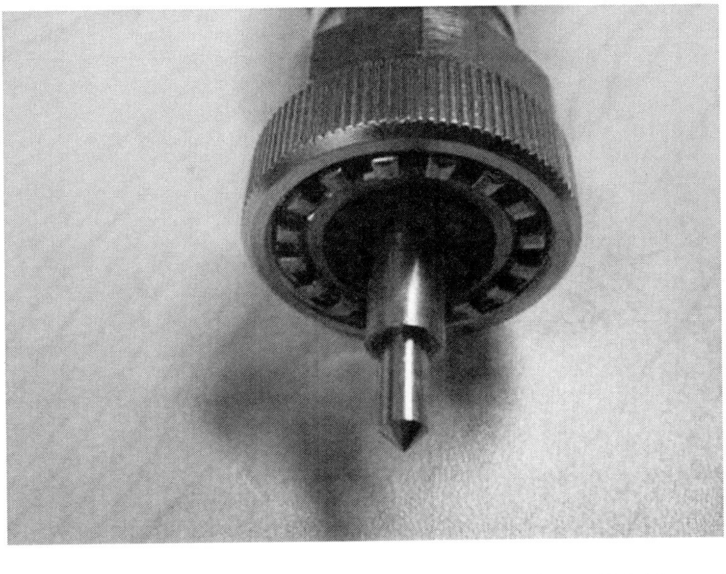

Figure 7. Test fixture and sample for permittivity from 50MHz to 5GHz.

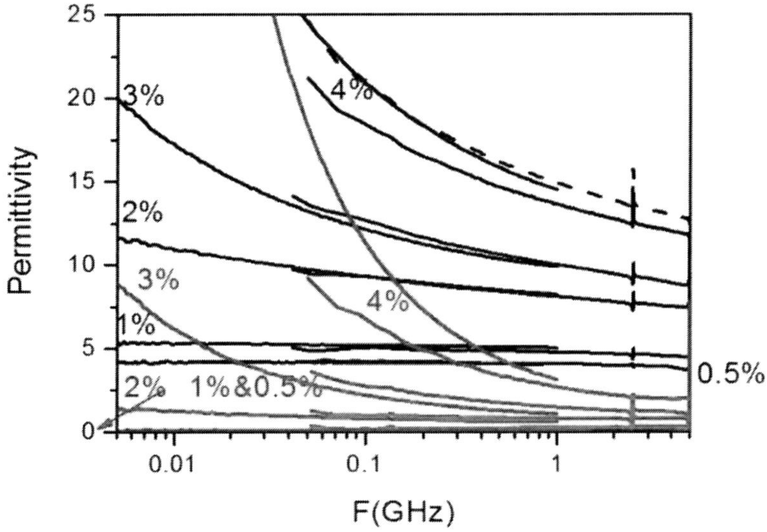

Figure 8. Permittivity of SWCNT composites measured by Impedance and 14mm coaxial method.

4. RESULTS AND DISCUSSION

4.1. Electrical Properties

The measurement electrical properties of different types of CNTs are plotted in Figure 9. The marks are the measured resistivity of CNTs composites prepared by sonication process (stars) and manual mixing (squares). And the poly-lines are the guide of the dependence of resistivity on the concentration of CNTs instead of fitting curves. Due to the measurement range of the insulation multimeter is up to 200G Ω, the resistibility measured at low concentration might be lower than the real value which could results in larger percolation threshold.

For SWCNT and DWCNT, the percolation threshold is around 1wt% if dispersed by sonication process. The percolation of manually mixed SWCNT and DWCNT composites could be up to 3~4wt%. For MWCNT composites, it seems the resistivity decreases before 1%. The percolation ratio could be between 0.5% and 1%. No obvious difference could be found from MWCNT composites dispersed by two different processing methods. Individual tubes instead of bundles could be observed from SEM image of MWCNT in Figure 1(c), as compared with SWCNT (Figure 1(a)) and DWCNT (Figure 1(b)). Therefore, manual mixing might be enough to disperse the tubes inside polymer matrix. However, sonication mixing could result in more steep percolation transition for SWCNT and DWCNT composites. And the bundle of SWCNT and DWCNT couldn't be broken by the manual mixing process.

4.2. Frequency-Dependent Permittivity

Figure 10 to Figure 12 show the measured dielectric properties of SWCNT, DWCNT and MWCNT composites processed with sonication method respectively. The percolation for three types of CNTs is around 1%. It is sufficient to investigate concentration up to 5% from percolation phenomena point of view. In all of three figures, solid lines show the measured real (in black) and imaginary (in red) permittivity and dash lines show fitting curved using (1). The permittivity of silicone host (ɜ *h*) is constant value, 3, according to impedance measurement. From the measured dependence of the permittivity, the dispersion curves for three types of CNTs appear to be similar and do not show any absorption peak that is typical of the Debye dispersion law and mixing rules. The frequency dispersion is smooth and resembles the dispersion of the percolation type. Therefore, the fitting of date made with scaling dispersion law instead of the Cole-Davidson law is validated (Liu et al, 2007b). The frequency dependence of the real and imaginary permttivity are fitted separatedly, to track deviations from the percolation law given by (2).

The parameters x and y obtained by curve fitting, which depend on the concentration, are shown inTable 2. Both A and B increase steadily with concentration which is understandable due to more conducting channels involved at higher concentrations. Therefore, they are not included in Table 2. Both x and y increase with filling factor. The critical exponent y is smaller than x for most of concentraions for all of three types of CNTs. Exponent x at percolation threshold fall in the the range (0.2 to 0.4) form theoretical calculation, but exponent y is relatively lower (0.1 to 0.2) for 2%. Deviation from the linear dependence (scaling law) can be found from 4% to 5% may explain the higher exponent value than the universal value.

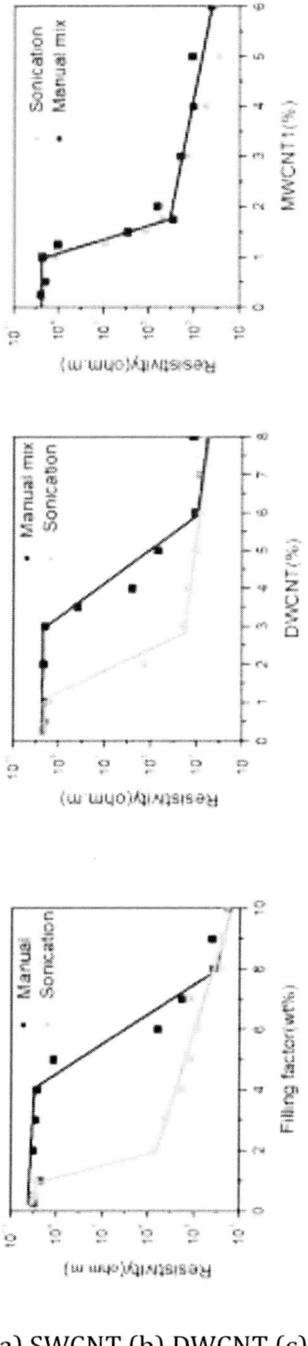

Figure 9. Resistivity of (a) SWCNT (b) DWCNT (c)MWCNT.

Therefore, the scaling frequency dispersion is observed in the CNTs composites within wide frequency range and concentration around percolation. In the percolation theory, the scaling behavior depends on the geometric spread of the properties of clusters, which is a function of the number of inclusions comprising a cluster. In actual composites, certain reasons lead to an additional spread of the properties of the clusters, such as the dispersion in dimension and in the conductivity of the CNT. Another reason can be the imperfect electrical contacts among the inclusion in the clusters where the conductivity between adjacent inclusions is due to interpaticle tunneling. An additional spread of the dispersion curve may be due to the elongated shape of CNT, because a net of tubes forms clusters of complex shape, which must be described by an assembly of RC circuits rather than by a single circuite. All these phenomana change the distibution of properties of tube clusters and therefore, results different frequency-dispersion from and exponents.

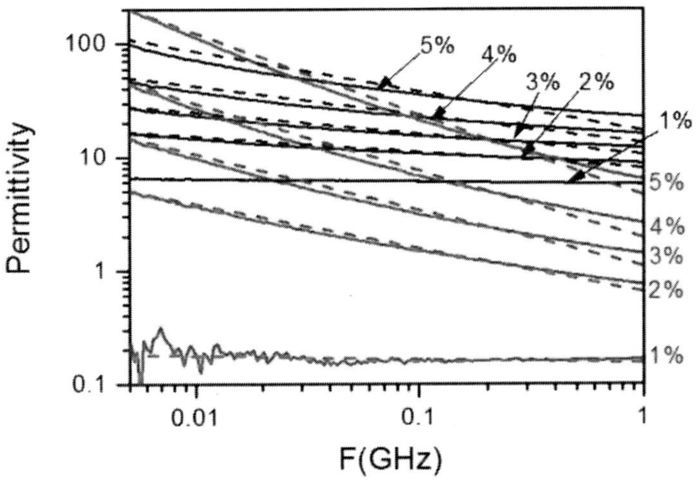

Figure 10. Permittivity of SWCNT composite.

4.3. Tunable Permittivity Under Bias Voltage

Figure 13 shows the dependence of normalized real and imaginary permittivity at 1 MHz, of the SWCNT composites at various concentrations, measured under a biasing voltage. The dependence of permittivity on bias voltage can be found at high frequency till a few hundred MHz. Figure 14 plots the tunability of permittivity of SWCNT composite from 1MHz to 100MHz. However, with increasing measurement frequency, the tunability becomes smaller. When the concentration is equal to or smaller than 2%, both real and imaginary permittivities do not change with voltage up to 15 V. It is understandable from the percolation threshold point of view because most of the tubes are out of the conducting network. Therefore, no voltage can be applied to isolate tubes or tube clusters. Once the concentration is higher than 3%, permittivity becomes sensitive to bias voltage. The tunability becomes more pronounced with increasing concentration. Samples with concentration of CNTs higher than 5% were not investigated due to extremely large imaginary permittivity. Obviously, bias voltage can decrease real permittivity but increase the imaginary part as shown in Figure 13. If the whole conducting network is treated as a circuit containing resistors and capacitors, replacing some capacitors with resistors will achieve the effect to reduce the real part and raise the imaginary part of permittivity. This is what was happening when bias voltage was applied to the samples in which junctions are formed by overlapping tubes. Electrons hop among tubes driven by bias voltages. The interaction among tubes becomes more resistive instead of capacitive under bias voltage. The tunability, defined as the change under bias over the original value, is more than 60% for the real permittivity and 80% for the imaginary permittivity for the sample with 5% of tubes. It is important to emphasize that the tunability of the CNT composites was reversible over a few thousand of cycles. Similar effect has also been found out from DWCNT and MWCNT composites when the concentration is large than percolation value within similar frequency range.

The novel tunable permittivity of the CNT composites may have potential applications in electromagnetic smart materials and structures. A smart materials or structures basically possess electrically or magnetically adjustable transmission or reflection coefficient to incident electromagnetic wave. The simplest solution is on-off switch, which was realized by pin diodes. One of the drawbacks of pin diodes is their relatively weak tolerance to high current, which is usually encountered in practical applications. A possible approach is to use a small piece of CNT composite to replace pin diode. Without the application of a bias voltage, the composite is insulating (off), while it becomes conducting under a certain bias voltage (on). Comparing with diodes, CNT composites can withstand higher current, voltage and power. Although the tunability of the SWCNT composite is still not high enough to provide similar on-off effect as a pin diode, we are confident that it can be further improved through proper dispersion or alignment of CNTs within holding matrix.

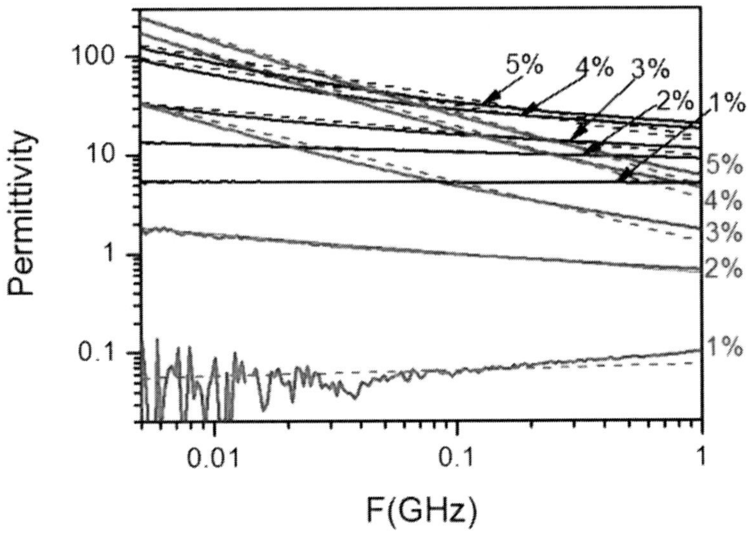

Figure 11. Permittivity of DWCNT composite.

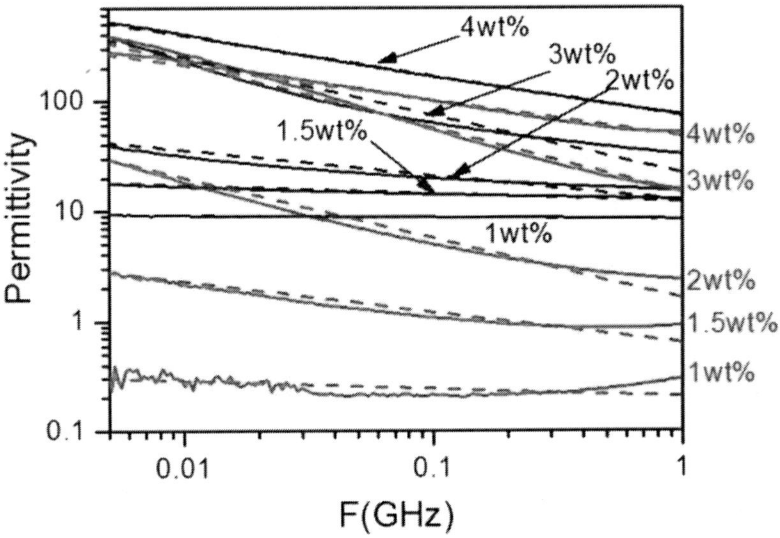

Figure 12. Permittivity of MWCNT composite.

Table 2. Critical exponents as function of concentration obtained by linear fitting of permittivity.

SWCNT			DWCNT			MWCNT		
wt%	y	x	wt%	y	x	wt%	y	x
1	0.04	0.03	1	0.02	-0.05	1	0.03	0.07
2	0.18	0.39	2	0.11	0.21	1.5	0.09	0.28
3	0.217	0.491	3	0.276	0.62	2	0.26	0.55
4	0.275	0.591	4	0.402	0.736	3	0.54	0.59
5	0.368	0.713	5	0.438	0.747	4	0.37	0.32

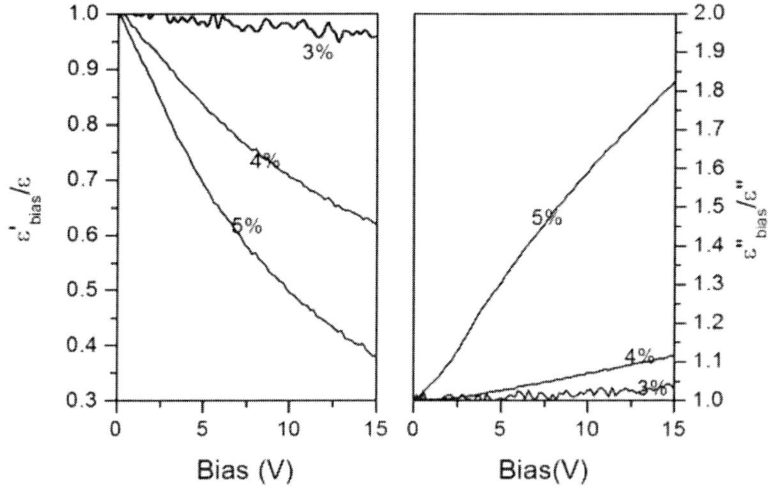

Figure 13. Tunable permittivity of SWCNT composites (5MHz).

5. CONCLUSIONS

In this chapter, the complex permittivity of SWCNT, DWCNT and MWCNT composites at microwave band is obtained experimentally by the impedance method. The frequency dependence of the permittivity is well described by the scaling law based on the percolation theory. CNT composite is a system comprising clusters and conducting networks, the microwave behavior of which depends on the cluster formation, instead of the properties of individual fibers. Strong frequency-dependent behavior may be observed in measurement due to size effects. The microwave properties of the composites are found to be very sensitive to the preparation method. However, certain deviations are observed in the scaling parameters derived from the percolation theory. Namely the critical exponents are different from conventionally accepted universal values. This may be attributed to the non-perfect electrical contacts within clusters, which typically are not considered in mixing laws.

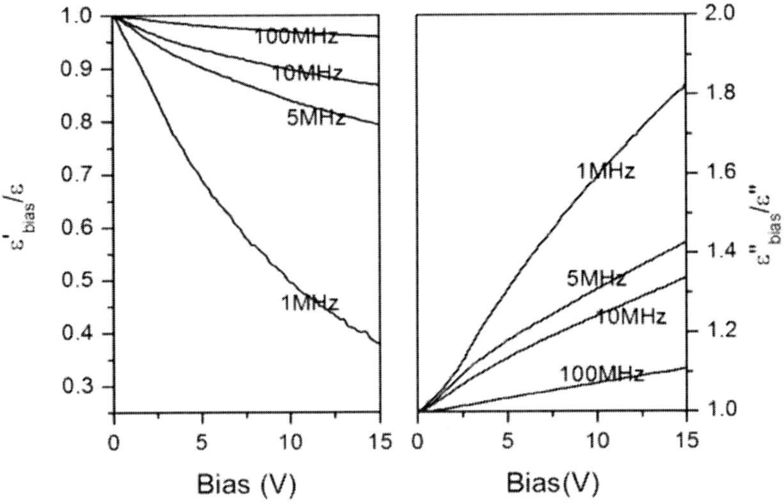

Figure 14. Tunable permittivity of SWCNT composites (5%).

The CNT composites have large real permittivity at microwaves, adjustable dielectric loss tangent, and noticeable frequency dispersion of permittivity, even at concentrations above the percolation threshold. For these reasons, CNT composites with $p>p_c$ are probably useful as microwave dielectrics and frequency dispersive materials. CNT composites have been found to possess novel tunable effective permittivity under small bias voltages. The tunable properties can be attributed to flowing of electrons in the conducted network formed by tubes and tube clusters. Hence they are only obtainable at concentrations above percolation threshold. The tunability of both real and imaginary permittivity has potential applications in smart material and structure, which is the research subject that is currently on-going in our group.

ACKNOWLEDGEMENTS

The authors appreciate the fruitful discussion with Dr. K. Rozanov (ITAE, Russia) and Dr. C. R. Deng (DSO, Singapore). The research was supported by Defence Science and Technology Agency (DSTA), Singapore, under project POD0103671. Wen-Yan Yin appreciates the financial support by the National Basic Research Program under Grant 2009CB320204, and the NSFC under Grant 60821062 of CHINA

REFERENCES

1. A. N. Lagarkov, S. M. Matitsine, K. N. Roznov, A. K. Sarychev, 1998 Dielectric properties of fiber-filled composites, J. Appl. Phys., 84 3806 14

2. A. K. Jonscher, 1999 Dielectric relaxation in solids, J. Phys. D- Appl. Phys. 32, n 14, 57 70 .

3. B. K. Kim, J. Lee, I. Yu, 2003 Electrical properties of single-wall carbon nanotube and epoxy composites, J. Appl. Phys., 94(10), 6724-8

4. C. A. Martin, J. K. W. Sandler, M. S. P. Shaffer, M. K. Schwarz, W. Bauhofer, K. Schulte, A. H. Windle, 2004 Formation of percolation networks in multi-wall carbon-nanotube-epoxy composites, Composites science and technologies, 64 2309 16

5. C. Brosseau, 2002 Generalized effective medium theory and dielectric relaxation in particle-filled polymeric resins, J. Appl. Phys. 91(5), 3197-204.

6. C. P. Neo, V. K. Varadan, 2004 Optimization of carbon fiber composite for microwave absorber, IEEE Trans. Electromagn. Compat. 46 102 6

7. C. R. Simovsky, S. A. Tretyakov, A. H. Sihvola, M. M. Popov, 2000 On the surface effect in thin molecular or composite layers, Eur. Phys. J. Appl. Phys. 9 195 204

8. C. Yang, Y. Lin, C. W. Nan, 2009 Modified carbon nanotube composites with high dielectric constant, low dielectric loss and large energy density (2009). Carbon, 47 1096 101

9. J. N. Coleman, S. Curran, A. B. Dalton, A. P. Davey, B. Mccarthy, W. Blau, R. C. Barklie, 1998 Percolation-dominated conductivity in a conjugated-polymer-carbon-nanotube composite, Phys Rev B, 58(12), 7492-5

10. D. Satuffer, A. Aharony, 1991 Introduction to Percolation Theory, Taylor & Francis, 2nd edition.

11. D. J. Bergman, Y. Imry, 1977 Critical behaviour of the complex dielectric constant near the percolation threshold of a heterogeneous material, Phys. Rev. Lett., 39 1222 25

12. F. Bordi, C. Cametti, J. Rouch, F. Sciortino, P. Tartaglia, 1996 Cluster formation in water-in-oil microemulsions at percolation: evaluation of the electrical properties, J. Phys.: Condens. Matter. 8(25), 19-37.

13. H. Y. Lee, Y. M. Shkel, 2007 Self-sensing and self-actuation response of carbon nanotube composites, Proc. Of SPIE, 6526 1 10 .

14. I. J. Youngs, A. S. Treen, G. Fixter, S. Holden, 2002 Design of solid broadband human tissue simulant materials, IEE Proc. Sci. Meas. Technol. 149 323 8 .

15. J. B. Donnet, R. C. Bansal, M. J. Wang, 1993 Carbon Black: Science and Technoloy, Dekker, New York

16. J. K. W. Sandler, J. E. Krik, I. A. Kinloch, M. S. P. Shaffer, A. H. Windle, 2003 Ultra-low electrical percolation threshold in carbon-nanotube-epoxy composites, Polymer, 44 5893 99

17. J.P. Straley 1977 Critical exponents for the conductivity of random resistor lattices, Phys, Rev. B, 15 5733 7 .

18. K. Park, S. Lee, C. Kim, J. Han, 2006 Fabrication and electromagnetic characteristics of electromagnetic wave absorbing sandwich structures, Composites science and technology, 66 576 84

19. L. Liu, L. B. Kong, S. Matitsine, 2008 Tunable effective permittivity of carbon nanotube composites, Appl. Phys. Lett., 93, 113106

20. L. Liu, S. Matitsine, Y. B. Gan, K. Rozanov, 2007a Cluster effect in frequency selective composites with randomly distributed long conductive fibres, J. Phys. D: Appl. Phys. 40 7534 9

21. L. Liu, S. Matitsine, Y. B. Gan, L. F. Chen, L. B. Kong, 2007b Frequency dependence of effective permittivity of carbon nanotube composites, J. Appl. Phys. 101, 094106.

22. Li-Li Wang, Beng-kang Tay, Kye-yak See, Zhuo Sun, Lin-Kin Tan, Darren Lua, 2009 Electromagnetic interference shielding effectiveness of carbon-based materials prepared by screen printing, Carbon, 47 1905 10

23. L. Wei, B. Wang, Q. Wang, L. Li, Y. Yang, Y. Chen, 2008 Effect of Centrifugation on the Purity of Single-Walled Carbon Nanotubes from MCM-41 Containing Cobalt, J. Phys. Chem., 112 17567 75

24. M. A. Seo, J. H. Yim, Y. H. Ahn, F. Rotermund, D. S. Kim, S. Lee, H. Lim, 2008 Terahertz electromagnetic interference shielding using single-walled carbon nanotube flexible films, Appl. Phys. Lett., 93, 231905

25. M. Meyyapan, 2005 Carbon Nanotubes: Science and Applications, CRC Press, LLC

26. M. H. Al-Saleh, Sundararaj. Uttandaraman, 2009 Electromagnetic interference shielding mechanisms of CNT/polymer composites, Carbon, 47 1738 46 .

27. P. C. P. Watts, D. R. Ponnampalam, W. K. Hsu, A. Barnes, B. Chambers, 2003 The complex permittivity of multi-walled carbon nanotube-polystyrene composite films in X-band, Chem. Phys. Lett., 378 609 14 .

28. R. B. Laibowitz, Y. Gefen, 1984 Dynamic scaling near the percolation threshold in thin Au films, Phys. Rev. Lett., 53 380 3

29. R. Saito, G. Dresselhaus, M. S. Dresselhaus, 1998 Physical Properties of Carbon Nanotubes, Imperial College Press, London.

30. S. Iijima, 1991 Nature Helical microtubules of graphitic carbon, 354, 56.

31. S. L. Browning, J. Jodge, R. R. Price, J. Schelleng, P. E. Schoen, D. Zabetakis, 1998 Fabrication and radio frequency characterization of

high dielectric loss tubule-based composites near percolation, J. Appl. Phys., 84 6109 13 .

32. S. Lee, J. Kang, G. Kim, 2006 Fabrication and design of multi-layered radar absorbing structures of MWCNT-filled glass/epoxy plain-weave composites, Composite structures, 76 397 405

33. S. Shi, J. Liang, 2008 the effect of multi-wall carbon nanotubes on electromagnetic interference shielding of ceramic composites, Nanotechnology, 19, 255707

34. S. M. Matitsine, K. M. Hock, L. Liu, Y. B. Gan, A. N. Lagarkov, K. N. Rozanov, 2003 Shift of resonance frequency of long conducting fibers embedded in a composite, J. Appl. Phys. 94 1146 54 .

35. T.D. Burchell 1999 Carbon Materials for Advanced Technologies, Elsevier Science Ltd, Oxford

36. Y. Gefen, A. Aharony, S. Alexander, 1983 Anomalous diffusion on percolating clusters Phys. Rev. Let., 50 1 77 80 .

37. Y. Huang, N. Li, Y. , F. Du, F. Li, X. He, X. Lin, H. Gao, Y. Chen, 2007 The influence of single-walled carbon nanotube structure on the electromagnetic interference shielding efficiency of its epoxy composites, Carbon, 45 1614 21

38. Y. Song, T. W. Noh, S. I. Lee, J. R. Gaines, 1983 Experimental study of the three-dimensional AC conductivity and dielectric constant of a conductor-insulator composite near the percolation threshold, Phys. Revs. B, 33(2), 904-8

10

QUANTUM-COUPLED RADIAL-BREATHING OSCILLATIONS IN DOUBLE-WALLED CARBON NANOTUBES

Kaihui Liu[1,2], Xiaoping Hong[1], Muhong Wu[2], Fajun Xiao[1], Wenlong Wang[2], Xuedong Bai[2], Joel W. Ager[3], Shaul Aloni[4], Alex Zettl[1,3], Enge Wang[5] & Feng Wang[1,3]

[1]Department of Physics, University of California at Berkeley, Berkeley, California 94720, USA.

[2]Institute of Physics, Chinese Academy of Sciences, Beijing 100190, China.

[3]Materials Sciences Division, Lawrence Berkeley National Laboratory, Berkeley, California 94720, USA.

[4]The Molecular Foundry, Lawrence Berkeley National Laboratory, Berkeley, California 94720, USA.

[5]International Center for Quantum Materials, School of Physics, Peking University, Beijing 100871, China

Van der Waals-coupled materials, ranging from multilayers of graphene and MoS_2 to superlattices of nanoparticles, exhibit rich emerging behaviour owing to quantum coupling between individual nanoscale constituents. Double-walled carbon nanotubes provide a model system for studying such quantum coupling mediated by van der Waals interactions, because each constituent single-walled nanotube can have distinctly different physical structures and electronic properties. Here we systematically investigate quantum-coupled radial-breathing mode oscillations in chirality-defined double-walled nanotubes by combining simultaneous structural, electronic and vibrational characterizations on the same individual nanotubes. We show that these radial-breathing oscillations are collective modes characterized by concerted inner- and outer-wall motions, and determine quantitatively the tube-dependent van der Waals potential governing their vibration frequencies. We also observe strong quantum interference between Raman scattering from the inner- and outer-wall excitation pathways, the relative phase of which reveals chirality-dependent excited-state potential energy surface displacement in different nanotubes.

Electronic and vibrational coupling in the nanoscale governs many important physical processes, such as energy transfer in photosynthesis[1], charge separation in heterojunctions[2] and thermal transport through interfaces[3]. Quantum coupling between nanostructures is also crucial for understanding emerging properties in van der Waals-coupled superstructures, such as multilayer of graphene and superlattice of nanoparticles. Double-walled carbon nanotubes (DWNTs), a coaxial composite of two single-walled carbon nanotubes (SWNTs), provide a unique model system for quantitative study of such nanoscale coupling, because the structural and physical properties of DWNTs are precisely defined individually, but richly varied among different species[4,5, 6, 7, 8, 9, 10, 11, 12]. Specifically, a SWNT is uniquely defined by its chiral indices (n, m). A DWNT, composed of two SWNTs, is fully defined by chiral indices $(n_o, m_o)/(n_i, m_i)$ of the coaxial outer-/inner-walls[10, 11, 12]. Its electronic and vibrational properties, however, can differ profoundly from those of the constituent SWNTs owing to tube–tube interactions.

Radial-breathing mode (RBM) oscillation is a signature vibration of one-dimensional nanostructures. It is widely used for characterizing carbon nanotubes[13, 14, 15, 16, 17], and can be sensitively measured at the single-nanotube level using resonance Raman spectroscopy[18, 19, 20]. In SWNTs, the RBM oscillation frequency (ω_{RBM}) uniquely determines the nanotube diameter (D) following a simple scaling law $\omega_{RBM}=228/D$ (nm cm^{-1})[21]. The RBM behaviour in DWNTs, however, is drastically different owing to strong inter-tube coupling. The coupling effects in DWNTs have been actively investigated for over one decade[4, 5, 6, 7, 8, 9], but they are still poorly understood. In this letter, we systematically study quantum-coupled RBM oscillations in chirality-defined DWNTs by simultaneously determining structural, electronic and vibrational properties of individual nanotubes with combined single-tube electron diffraction[11, 12], Rayleigh scattering[22, 23]and Raman scattering[18, 19, 20] techniques.

We show that mechanical coupling between the walls lead to collective oscillation modes with concerted inner- and outer-wall vibrations in DWNTs. Oscillation frequencies of the coupled RBM modes can be quantitatively described by a coupled oscillator model, where the coupling originates from the van der Waals force between the inner and outer tubes. Different DWNTs, which have different inner- and outer-wall separations, allow us to probe the inter-tube van der Waals interactions at an effective pressure that can be both positive and negative and at gigapascal level. The coupled RBM oscillations also exhibit unusual electron–phonon interactions. They always appear in pairs in resonance Raman spectra, because each RBM mode couples to electronic resonances in both walls. Furthermore, Raman amplitudes from inner- and outer-wall excitation channels can interfere quantum mechanically. Relative interference phase between the two Raman channels enabled us to examine the displaced excited-state potential energy surface in SWNTs and its dependence on the nanotube chirality.

1. RESULTS

1.1 Experimental design

Figure 1 shows the schematic illustration of our experimental design. We utilized the SiO_2/Siwafer with open slit as our base substrate, where suspended carbon nanotubes are directly grown on (See Methods for more details). Transmission electron microscope (TEM) beam and laser beam can both go through the slit. This design enables the combination of TEM electron diffraction, Rayleigh scattering and Raman scattering techniques to probe the structural, electronic and vibrational properties of the same individual nanotubes. In addition, the use of the as-grown suspended nanotubes makes to exclude the substrate effects and ensures to probe the intrinsic physical properties of carbon nanotubes.

The substrate is etched with open slit, where individual suspended nanotubes are directly grown on. TEM beam and laser beam can both go through the slit. This design enables the combination of TEM electron diffraction, Rayleigh scattering and Raman scattering techniques to, respectively, probe the structural, electronic and vibrational properties of the same individual carbon nanotubes.

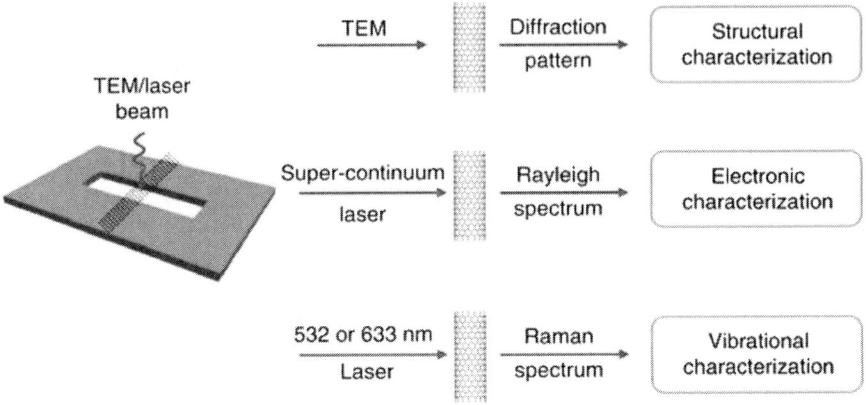

Figure 1. Schematic illustration of the experimental design.

1.2 Coupled RBM vibrations in DWNTs

Figure 2 displays electron diffraction patterns, Rayleigh scattering and Raman scattering spectra for two representative individual DWNTs. Electron diffraction patterns (Fig. 2a) unambiguously determine the DWNT chiral indices $(n_o, m_o)/(n_i, m_i)$ to be (27, 5)/(18, 5) and (31, 4)/(15, 13). Electronic transitions of the two DWNTs are obtained from the Rayleigh scattering spectra (insets in Fig. 2b, respectively), in which prominent optical resonances can be observed. By comparing the optical resonances in Rayleigh spectra of DWNTs to that of the constituent SWNTs, we can assign the resonance peaks to inter-subband electronic transitions of either inner- and outer-wall nanotubes because tube–tube interactions in DWNTs produces only a small shift (tens of meV) of the electronic transition energies[24]. We measure the RBM oscillations of DWNTs using resonant Raman scattering. Two well-defined RBM peaks are observed in the resonant Raman spectra of the (27, 5)/(18, 5) and (31, 4)/(15, 13) DWNTs (Fig. 2b). The laser excitation energy (1.96 eV) is marked by dashed line in the figure insets showing the Rayleigh scattering spectra. We have studied 35 DWNTs with well-defined Rayleigh scattering resonances (that can be assigned to each tube components) and TEM diffraction data. Of these DWNTs, 13 have one or more electronic transitions in resonance with our Raman excitation lasers. RBM Raman peaks are observed in, and only in, these 13 tubes. We summarized in Table 1 their chirality and Raman data. Our DWNTs have relatively large diameters, with inner-wall diameter (D_i) >1.5 nm in all the nanotubes.

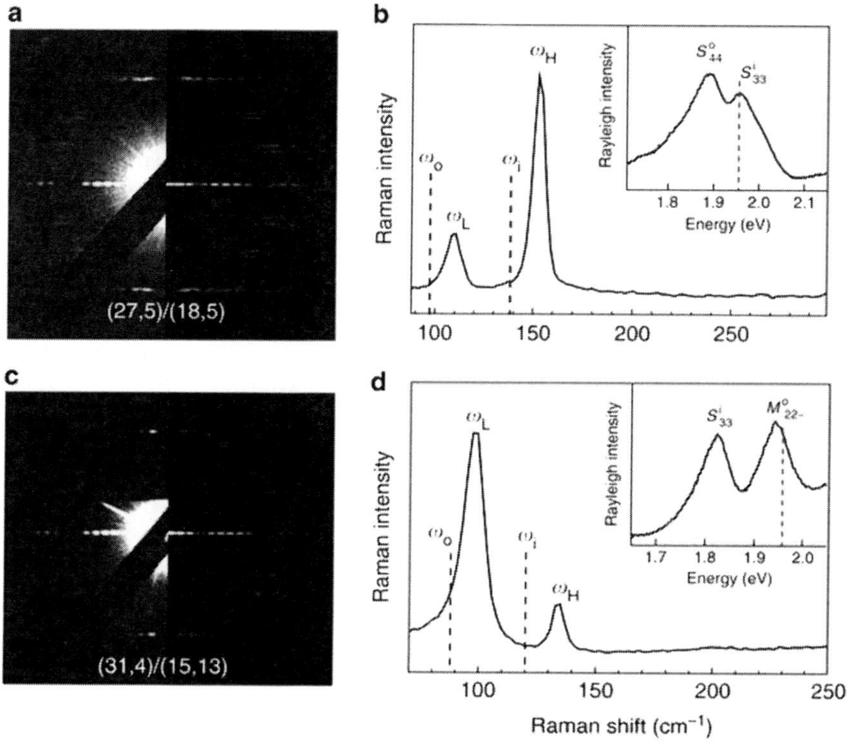

Figure 2. Coupled radial-breathing mode vibrations in double-walled carbon nanotubes. (**a,b**) TEM electron diffraction pattern and Raman spectrum for DWNT (27, 5)/(18, 5). Two well-defined RBM peaks are present in the Raman spectrum, and their frequencies are strongly blue-shifted from those of the constituent single-walled nanotubes (indicated by dashed lines). The excitation photon energy of 1.96 eV is in resonance with both inner-wall (S^i_{33}) and outer-wall (S^o_{44}) electronic transitions, as shown in the Rayleigh scattering spectrum (inset of **b**). (**c,d**) TEM electron diffraction, Raman and Rayleigh spectrum data for the (31, 4)/(15, 13) DWNT. Two appreciably blue-shifted RBM oscillations are again observed. In this DWNT, Raman excitation energy (1.96 eV) is only in resonance with the M^o_{22-} transition of the metallic outer-wall nanotube.

There are two striking features in the RBM Raman data of these (relatively large diameter) DWNTs. (1) Frequencies of the two RBM oscillations in DWNTs, denoted as ω_L and ω_H, are respectively much higher than the RBM frequencies of the constituent outer-(ω_o) and inner-wall SWNTs (ω_i) (for example, Fig. 2b). If this blue shift is attributed to a simple increase of 'effective' restoration force, this restoration force has to increase by as high as 35%. (2) The resonant peaks almost always appear in pairs in resonant Raman spectra: there are either no RBM peak (in 21 tubes) or two RBM peaks (in 13 tubes). Only one nanotube has just one RBM peak. This behaviour is quite surprising, because conventional picture suggests that resonance enhancement of the inner- and outer-wall RBM oscillations are largely independent[5, 7] and most nanotubes should have a single RBM peak when excited resonantly. To understand the unusual RBM behaviour in DWNTs, we have to realize that tube–tube interactions in these DWNTs are not small perturbations. Instead, they qualitatively change the DWNT RBM oscillations by strongly mixing the inner- and outer-wall vibrations: the observed lower-(ω_L) and higher-frequency (ω_H) RBM modes are collective oscillations with concerted in-phase and out-of-phase motion of the two walls.

1.3 Inter-wall van der Waals interactions in DWNTs

The collective RBM oscillations of a unit-length DWNT can be modelled by a coupled mechanical oscillator with five parameters: m_i, k_i, m_o, k_o and k_c (Fig. 3a)[25]. The unit-length mass of the inner- and outer-walls (m_i and m_o) can be obtained directly from the tube chiral indices, and the unit-length intrinsic force constant of the two walls (k_i and k_o) can be obtained from RBM frequencies of isolated SWNTs (Supplementary Note S1). The only unknown parameter is the coupling force constant k_c, which characterizes the van der Waals interaction between the two walls of the DWNT. Using a single value of k_c for any given DWNT, we can simultaneously reproduce the in-phase (ω_L) and out-of-phase RBM vibration frequencies (ω_H) (Fig. 3b). For different DWNTs, the coupling constant k_c vary significantly.

Table 1. Chiral indices and coupled radial-breathing mode oscillation data for double-walled carbon nanotubes.

Sample	(n_o, m_o)	(n_i, m_i)	D_o	D_i	ω_L	ω_H	Resonance	s	$(I_L/I_H)_{exp}$	$(I_L/I_H)_{N-I}$	$(I_L/I_H)_I$
1	(40, 1)	(22, 14)	3.17	2.46	79	115	Outer	−1	19	5.8	20
2	(31, 4)	(15, 13)	2.60	1.90	98	137	Outer	−1	8.0	10	8.2
3	(27, 5)	(18, 5)	2.33	1.64	108	155	Outer	−1	24	23	25
4	(32, 11)	(23, 10)	3.03	2.30	83	112	Inner	+1	0.17	0.18	0.16
5	(26, 17)	(27, 2)	2.94	2.20	84	114	Inner	+1	0.08	0.12	0.09
6	(21, 21)	(26, 2)	2.85	2.12	85	120	Inner	+1	0.37	0.16	0.33
7	(20, 19)	(24, 1)	2.64	1.92	93	129	Inner	−1	0.17	0.10	0.13
8	(19, 15)	(19, 3)	2.31	1.62	110	156	Inner	−1	0.10	0.15	0.11
9	(27, 12)	(20, 9)	2.71	2.01	93	133	Both	−1	3.0	1.2	2.9
10	(23, 13)	(14, 12)	2.47	1.76	103	143	Both	+1	5.0	2.2	4.4
11	(27, 6)	(15, 10)	2.38	1.71	103	154	Both	−1	1.5	2.8	1.2
12	(27, 6)	(15, 10)	2.38	1.71	105	155	Both	−1	0.83	0.49	0.86
13	(27, 5)	(18, 5)	2.33	1.64	108	154	Both	−1	0.37	0.16	0.42

Diametres are listed in unit of nm and frequencies in unit of cm^{-1}. Subscript exp, N − I, I denotes experimental value, prediction not including quantum interference and prediction with quantum interference, respectively.

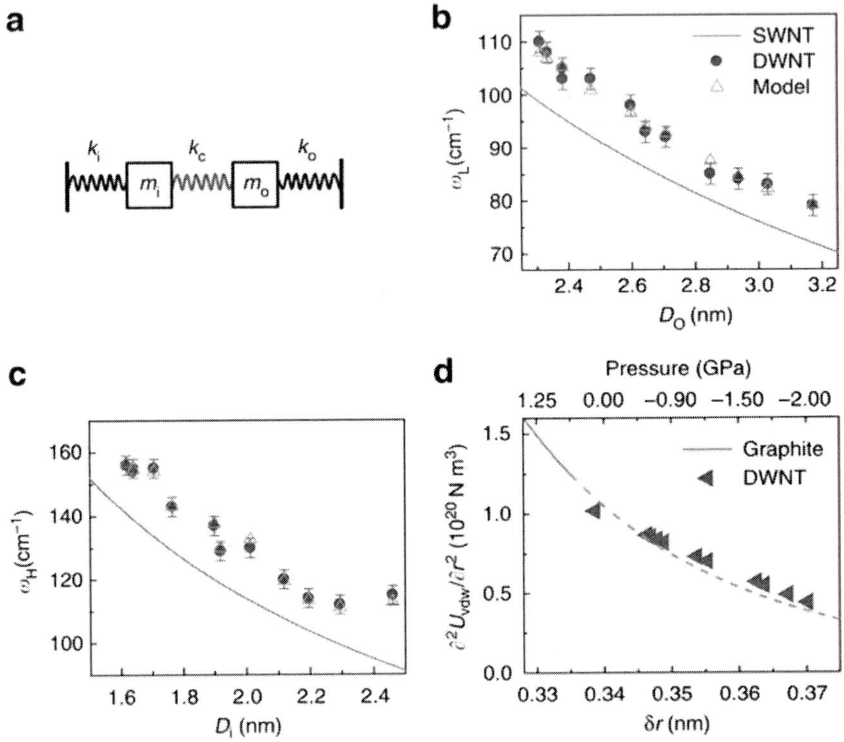

Figure 3. Coupled nanomechanical oscillations in double-walled carbon nanotubes. (**a**) Illustration of the coupled oscillator model, where m_i (m_o) and k_i (k_o) are, respectively, the inner (outer)-wall nanotube unit-length mass and intrinsic force constant, and k_c is the unit-length coupling force constant owing to tube–tube interactions. (**b**) Lower RBM frequencies of the DWNTs versus outer-wall nanotube diameters (dots), which are appreciably higher than uncoupled out-wall nanotube RBM frequencies (line). Theoretical prediction based on the coupled oscillator model (triangles) agrees well with our data. (**c**) Higher RBM frequencies of the DWNTs versus inner-wall nanotube diameters, which display analogous behaviour as in (**b**). The error bar in (**b**) and (**c**) represents the uncertainty in fitting the RBM resonance in Raman spectra. (**d**) Derived unit-area force constant owing to tube–tube van der Waals interaction, $\partial^2 U_{vdw}/\partial r^2$, for different inter-tube separation (triangles), and the comparison to results from high-pressure graphite measurements (line). The different inter-tube separation in DWNTs corresponds an effective pressure variation from 0.2 to −2.0 GPa (top label). (The zero-pressure inter-tube distance for DWNTs is slightly shifted from that of graphite owing to the finite tube curvature[4]).

The tube-dependent k_c provides a unique way to examine how the van der Waals interaction varies with the separation between the inner- and outer-wall tubes. The unit-length coupling force constant kc can be expressed approximately using the average unit-area inter-tube van der Waals potential U_{vdw} with $k_c = (\partial^2 U_{vdw}/\partial r^2) \cdot \pi \bar{D}$, where the mean diametre $\bar{D} = (D_i + D_o)/2$. In our as-grown DWNTs, the separation between the inner- and outer-tubes$\delta r = (D_o - D_i)/2$ can vary from 0.34 to 0.37 nm, which is uniquely determined by the tube chiral indices. Figure 3d displays the data for unit-area force constant *part;*$^2U_{vdw}$*/part;r*^{2}, which changes by over two times over the range of inter-tube separation existing in our DWNTs. Our results can be compared with the van der Waals interaction between unit-area graphene sheets under pressure obtained from compressibility measurements of graphite (solid line) and theoretical extrapolation in the negative pressure range (dashed line)[26], and they agree well. This comparison shows that the effective pressure between the walls of as-grown DWNTs reaches gigapascals owing to variations in the tube–tube separation[26, 27]. It is interesting to note that DWNTs allow us to readily access a negative pressure in the gigapascal range, which is difficult to achieve using conventional approaches. The force constant deduced here is related to the C_{33}modulus of graphite, and its value is about one order of magnitude larger than that from the shear mode of few-layered graphene, which is related to the C_{44} modulus of graphite[28, 29].

1.4 Quantum interference in resonance Raman process in DWNTs

Unlike separated inner- and outer-SWNTs RBM excitations (illustrated in Fig. 4a), the collective DWNT oscillations ω_L and ω_H contain both inner- and outer-wall motion, and they couple simultaneously to electronic transitions in both walls (Fig. 4b). Therefore, both coupled RBM oscillations will be resonantly excited if an electronic transition of either wall matches the excitation photon energy. This leads to the unusual behaviour that we observed experimentally: the RBM Raman modes are mostly either not observable (in 21 tubes) or appearing in pairs (in 13 tubes) in resonant Raman scattering spectra of DWNTs.

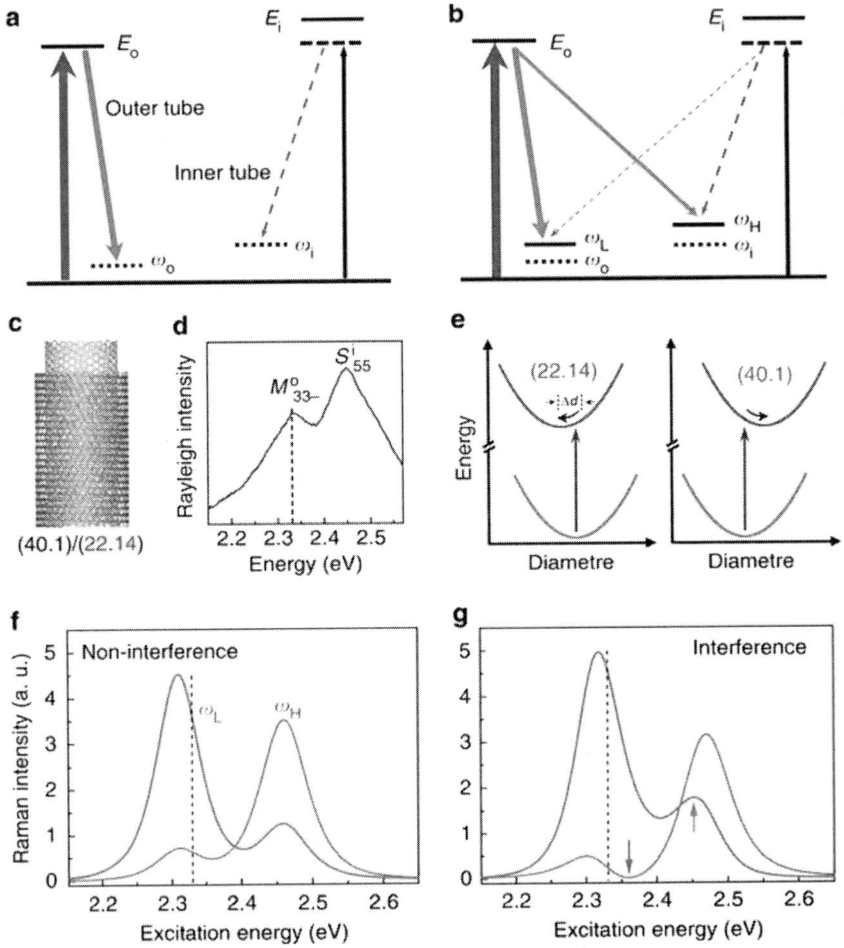

Figure 4. Quantum interference in resonance Raman process in double-walled carbonnanotubes. (**a**) Raman process with non-coupled outer and inner SWNT oscillations. The radial breathing mode (RBM) vibration of the SWNT exhibits strong Raman scattering when electronic transition resonant with excitation photon energy. Therefore, resonance Raman peaks for outer- and inner-SWNT appear independently. (**b**) Raman process with collective outer- and inner-wall oscillations. The coupled oscillations have higher frequencies, and each of them couples to both inner- and outer-SWNT electronic transitions. Resonance Raman peaks always appear in pairs even when only one electronic transition is in resonance. In addition, the vibrations from the inner- and outer-wall excitation channels will interfere with each other quantum mechanically. (**c**) Atomic model of DWNT (40, 1)/(22, 14). (**d**) Rayleigh spectrum of DWNT (40, 1)/(22, 14). The 2.33 eV Raman photon

excitation is resonant with the outer-wall metallic M_{33}^{o} transition but not the inner-wall semiconducting S_{55}^{i} transition, therefore two walls have opposite resonance phase factors. (**e**) The potential energy surfaces for the S_{55}^{i} and M_{33}^{o} excited states in the constituent (22, 14) and (40, 1) SWNTs. The displacements (Δd) are extremely small at ~1 fm, and they are exaggerated in the figure for illustration. The photo-induced RBM vibration can be understood in the Frank–Condon picture. Upon optical excitation the SWNT lattice, starting from the equilibrium configuration of the ground state, relaxes based on the excited-state potential energy surface and sets off the RBM vibration. (**f,g**) Simulated Raman intensities of out-of-phase (ω_H) and in-phase (ω_L) RBM modes versus excitation laser energy for the (40, 1)/(22, 14) DWNT without (**f**) and with (**g**) the quantum interference. The opposite resonance phase factors of the two walls (**d**), together with the oppositely displaced excited potential energy surfaces (**e**), will drive the inner- and outer-walls to vibrate mostly in the same direction after the optical excitation, and enhance the in-phase oscillation ω_L while suppress the out-of-phase oscillation ω_H owing to quantum interference (**g**). It leads to significantly enhance I_L/I_H at our Raman excitation energy 2.33 eV (dashed line), as observed experimentally.

The collective RBM oscillations in DWNTs couple to the electronic excitations quantum mechanically, and it leads to interesting quantum interference between the inner- and outer-wall excitation pathways. With such quantum interference even non-resonant contribution can become important, and Raman intensities of the coupled RBM oscillations in DWTNs are described by the superposition of Raman amplitudes from inner- and outer-wall excitation pathways as

$$I_L = |M_i R_i \langle \omega_i | \omega_L \rangle + M_o R_o \langle \omega_o | \omega_L \rangle|^2 \text{ and}$$
$$I_H = |M_i R_i \langle \omega_i | \omega_H \rangle + M_o R_o \langle \omega_o | \omega_H \rangle|^2$$

Here, $\langle \omega_{i(o)} | \omega_{L(H)} \rangle$ is the inner(outer)-wall component of the coupled RBM mode $\omega_L(\omega_H)$, and its value can be calculated directly from a quantized coupled mechanical oscillator model illustrated in Fig. 3a (Supplementary Note S2). $R_{i(o)}$ and $M_{i(o)}$ denotes, respectively, the electronic resonance factor and Raman matrix element of the inner(outer)-wall SWNT excitations. The resonance factor $R_{i(o)}$ can be obtained from Rayleigh scattering spectra of the DWNTs, which probe directly the optical resonances of both the inner- and outer-wall nanotubes (Supplementary Note S3). The Raman matrix element $M_{i(o)}$ characterizes how the phonon couples to the photo-excited exciton in the inner(outer)-wall nanotube. For simplicity, we will assume that M_i and $M_o$have the same magnitude and focus on their relative sign, s=sgn(M_i/M_o). This simplification emphasizes the importance of the phase factor, which determines whether the quantum interference is constructive or destructive for the coupled RBM oscillations.

To compare with the experiment quantitatively, we use the Raman intensity ratio I_L/I_H because it does not depend on an absolute determination of Raman cross-section. The ratio has the form of $I_L/I_H = \left|\frac{s \cdot R_i \langle \omega_i | \omega_L \rangle + R_o \langle \omega_o | \omega_L \rangle}{s \cdot R_i \langle \omega_i | \omega_H \rangle + R_o \langle \omega_o | \omega_H \rangle}\right|^2$ in our approximation. In Table 1, we listed the measured Raman intensity ratio $(I_L/I_H)_{exp}$ together with calculation results including the quantum interference $(I_L/I_H)_I$ and not including the interference $(I_L/I_H)_{N-I}$. Nice agreement between the theory and experiment is achieved when, and only when, the quantum interference effect is fully included using the correct s parameter.

We take a closer look at the Raman quantum interference using (40, 1)/(22, 14) DWNT as an example (Fig. 4c). The s parameter, which characterizes the relative sign between Raman matrix elements M_i and M_o, depends sensitively on the chirality of the inner- and outer-SWNTs. Its value is related to the relative motion direction of the inner- and outer-wall nanotubes immediately after the optical excitation. A physical picture of this process in outer- and inner-SWNTs is provided by recalling Frank–Condon effects (Fig. 4e): the excited state has a displaced potential energy surface compared with that of the ground state. The displacement is extremely small (~1 fm), but is essential for the exciton–phonon coupling. Upon optical excitation, the SWNT lattice, initially at the ground-state equilibrium configuration, relaxes based on the excited-state potential energy surface and sets off the RBM vibration. In (40, 1)/(22, 14) DWNT the potential energy surface displacements of inner- and outer-walls displace are opposite with an s=−1 (refs [30,31]30,31).

To determine the RBM oscillation interference, we also need to include the resonance phase factor. For (40, 1)/(22, 14) DWNT, the phase factors are nearly opposite for the (non-resonant) inner- and (resonant) out-wall pathways with 2.33 eV photon excitation (Fig. 4d). The opposite resonance phase factors, together with s=−1, determine that the quantum interference will suppress the Raman intensity of ω_H while enhance that of ω_L. This quantum interference leads to increased I_L/I_H ratio from 5.8 to 20, and the experimental value of I_L/I_H is 19 (Table 1).

We further plot the calculated RBM Raman intensity of the (40, 1)/(22, 14) DWNT at different excitation energies without (Fig. 4f) and with the quantum interference (Fig. 4g), which displays the variation of quantum interference effects with the excitation laser energy.

2. DISCUSSION

Our observed *s* parameter shows that the relative displacement of the excited-state potential energy surface varies with the SWNT chirality, and it can be described by a simple family pattern: the excited-state potential energy surface is displaced inwards for transitions on one side of the zig–zag cut through K point of the graphene Brillouin zone, which includes even transitions of mod($n-m$,3)=1 semiconducting, odd transitions of mod($n-m$,3)=2 semiconducting and higher branch transitions of non-armchair metallic nanotubes. The displacement is outwards for all other transitions. This is the first observation of family behaviour in the excited-state potential energy surface displacement, which has been predicted previously theoretically in *ab initio* and tight-binding calculations[30, 31].

The ability to combine structural, electronic and vibrational characterization on the same individual DWNTs allows us to systematically investigate quantum-coupled mechanical oscillations and their unusual interactions with electronic excitations for the first time. We show that van der Waals interaction, usually treated as a weak perturbation, can actually produce quantum phenomena in coupled nanostructures that have only been observed in covalent material previously. With exact structural determination, we are able to determine the separation-dependent van der Waals coupling between nanotube walls and the excited-state potential energy surfaces of constituent nanotubes. Such approach can also be used to explore other composite nanostructures. It will enable quantitative understanding of van der Waals interactions and the electronic and vibrational couplings in the nanoscale.

3. METHODS

In our experiment, we use suspended DWNTs that are free of substrate effects and compatible with both TEM and single-tube optical spectroscopy techniques. Long suspended nanotubes were grown by chemical vapour deposition (CVD) across open slit structures (~30 × 500 μm) fabricated on silicon substrates[32]. We use Fe nanoparticles as catalysts and methane in hydrogen(CH_4:H_2=1:2) as gas feedstock, and control the size of catalyst particles to achieve the selective abundance of DWNTs[33]. We determine chiral structures of DWNTs using electron diffraction with nano-focused 80 keV electron beams in a JEOL 2100 TEM[11, 12]. Utilizing the slit edges as markers, the same individual nanotubes can be identified in an optical microscopy setup. To determine electronic transitions of these individual nanotubes, we use Rayleigh scattering spectroscopy with a fibre-laser-based supercontinuum light source covering the spectral range from 450 to 900 nm (refs [22, 23]22,23). To determine RBM vibration and electron–phonon coupling in these individual nanotubes, we use resonant Raman scattering spectroscopy with laser excitation at photon energy of 1.96 or 2.33 eV (refs [18, 19, 20]18–20). Our instrumental resolution for RBM Raman frequency is ~2 cm^{-1}.

REFERENCES

1. Engel, G. S. *et al.* Evidence for wavelike energy transfer through quantum coherence in photosynthetic systems. *Nature* **446**, 782–786 (2007).

2. Granstrom, M. *et al.* Laminated fabrication of polymeric photovoltaic diodes. *Nature* **395**,257–260 (1998).

3. Wang, Z. H. *et al.* Ultrafast flash thermal conductance of molecular chains. *Science* **317**,787–790 (2007).

4. Saito, R. *et al.* Anomalous potential barrier of double-wall carbon nanotube. *Chem. Phys. Lett.* **348**, 187–193 (2001).

5. Bandow, S. *et al.* Raman scattering study of double-wall carbon nanotubes derived from the chains of fullerenes in single-wall carbon nanotubes. *Chem. Phys. Lett.* **337**, 48–54(2001).

6. Kociak, M. *et al.* Linking chiral indices and transport properties of double-walled carbon nanotubes. *Phys. Rev. Lett.* **89**, 155501–155504 (2002).

7. Villalpando-Paez, F. *et al.* Raman spectroscopy study of isolated double-walled carbon nanotubes with different metallic and semiconducting configurations. *Nano Lett.* **8**,3879–3886 (2008).

8. Liu, K. H. *et al.* Chirality-dependent transport properties of double-walled nanotubes measured *in situ* on their field-effect transistors. *J. Am. Chem. Soc.* **131**, 62–63 (2009).

9. Levshov, D. *et al.* Experimental evidence of a mechanical coupling between layers in an individual double-walled carbon nanotube. *Nano Lett.* **11**, 4800–4804 (2011).

10. Zuo, J. M. *et al.* Atomic resolution imaging of a carbon nanotube from diffraction intensities. *Science* **300**, 1419–1421 (2003).

11. Hirahara, K. *et al.* Chirality correlation in double-wall carbon nanotubes as studied by electron diffraction. *Phys. Rev. B* **73**, 195420–195430 (2006).

12. Liu, K. H. *et al.* Direct determination of atomic structure of large-indexed carbon nanotubes by electron diffraction: application to double-walled nanotubes. *J. Phys. D-Appl. Phys.* **42**,125412 (2009).

13. Saito, R., Dresselhaus, G. & Dresselhaus, M. S.. *Physical Properties of Carbon Nanotubes*Imperial College Press (1998).

14. Reich, S., Thomsen, C. & Maultzsch, J.. *Carbon Nanotubes: Basic Concepts and Physical Properties* Wiley-VCH, (2004).

15. Kurti, J., Kresse, G. & Kuzmany, H.. First-principles calculations of the radial breathing mode of single-wall carbon nanotubes. *Phys. Rev. B* **58**, R8869–R8872 (1998).

16. Sanchez-Portal, D. *et al.* Ab initio structural, elastic, and vibrational properties of carbon nanotubes. *Phys. Rev. B* **59**, 12678–12688 (1999).

17. Popov, V. N. & Henrard, L.. Breathinglike phonon modes of multiwalled carbon nanotubes.*Phys. Rev. B* **65**, 235415 (2002).

18. Fantini, C. *et al.* Optical transition energies for carbon nanotubes from resonant Raman spectroscopy: environment and temperature effects. *Phys. Rev. Lett.* **93**, 147406–147409(2004).

19. Telg, H. *et al.* Chirality distribution and transition energies of carbon nanotubes. *Phys. Rev. Lett.* **93**, 177401–177404 (2004).

20. Meyer, J. C. *et al.* Raman modes of index-identified freestanding single-walled carbon nanotubes. *Phys. Rev. Lett.* **95**, 217401–217404 (2005).

21. Liu, K. H. *et al.* Intrinsic radial breathing oscillation in suspended single-walled carbon nanotubes. *Phys. Rev. B* **83**, 113404 (2011).

22. Sfeir, M. Y. *et al.* Probing electronic transitions in individual carbon nanotubes by Rayleigh scattering. *Science* **306**, 1540–1543 (2004).

23. Sfeir, M. Y. *et al.* Optical spectroscopy of individual single-walled carbon nanotubes of defined chiral structure. *Science* **312**, 554–556 (2006).

24. Wang, F. *et al.* Interactions between individual carbon nanotubes studied by Rayleigh scattering spectroscopy. *Phys. Rev. Lett.* **96**, 167401 (2006).

25. Dobardzic, E. *et al.* The radial breathing mode frequency in double-walled carbon nanotubes: an analytical approximation. *Phys. Status Solidi. B* **237**, R7–R10 (2003).

26. Hanfland, M., Beister, H. & Syassen, K.. Graphite under pressure—equation of state and 1st-order raman modes. *Phys. Rev. B* **39**, 12598–12603 (1989).

27. Popescu, A., Woods, L. M. & Bondarev, I. V.. Chirality dependent carbon nanotube interactions. *Phys. Rev. B* **83**, 081406 (2011).

28. Tan, P. H. *et al.* The shear mode of multilayer graphene. *Nat. Mater.* **11**, 294–300 (2012).

29. Bosak, A. *et al.* Elasticity of single-crystalline graphite: inelastic X-ray scattering study.*Phys. Rev. B* **75**, 153408 (2007).

30. Machon, M. *et al.* Strength of radial breathing mode in single-walled carbon nanotubes.*Phys. Rev. B* **71**, 035416 (2005).

31. Jiang, J. *et al.* Exciton-photon, exciton-phonon matrix elements, and resonant Raman intensity of single-wall carbon nanotubes. *Phys. Rev. B* **75**, 035405 (2007).

32. Seidel, H., Csepregi, L., Heuberger, A. & Baumgartel, H.. Anisotropic etching of crystalline silicon in alkaline-solutions. 1. Orientation dependence and behavior of passivation layers.*J. Electrochem. Soc.* **137**, 3612–3626 (1990).

33. Huang, S. M., Cai, X. Y. & Liu, J.. Growth of millimeter-long and horizontally aligned single-walled carbon nanotubes on flat substrates.*J. Am. Chem. Soc.* **125**, 5636–5637 (2003).

11

CARBON NANOTUBES PRODUCED FROM AMBIENT CARBON DIOXIDE FOR ENVIRONMENTALLY SUSTAINABLE LITHIUM-ION AND SODIUM-ION BATTERY ANODES

Stuart Licht[*†], Anna Douglas[‡], Jiawen Ren[†], Rachel Carter[‡], Matthew Lefler[†], and Cary L. Pint[*‡]

[†] Department of Chemistry, George Washington University, Washington, D.C. 20052, United States

[‡] Department of Mechanical Engineering, Interdisciplinary Materials Science Program, and Vanderbilt Institute of Nanoscale Science and Engineering, Vanderbilt University, Nashville, Tennessee 37235, United States

ABSTRACT

The cost and practicality of greenhouse gas removal processes, which are critical for environmental sustainability, pivot on high-value secondary applications derived from carbon capture and conversion techniques. Using the solar thermal electrochemical process (STEP), ambient CO_2 captured in molten lithiated carbonates leads to the production of carbon nanofibers (CNFs) and carbon nanotubes (CNTs) at high yield through electrolysis using inexpensive steel electrodes. These low-cost CO_2-derived CNTs and CNFs are demonstrated as high performance energy storage materials in both lithium-ion and sodium-ion batteries. Owing to synthetic control of sp^3 content in the synthesized nanostructures, optimized storage capacities are measured over 370 mAh g^{-1} (lithium) and 130 mAh g^{-1} (sodium) with no capacity fade under durability tests up to 200 and 600 cycles, respectively. This work demonstrates that ambient CO_2, considered as an environmental pollutant, can be attributed economic value in grid-scale and portable energy storage systems with STEP scale-up practicality in the context of combined cycle natural gas electric power generation.

1. INTRODUCTION

A key challenge for atmospheric carbon capture and conversion technologies is the cost of operation or materials versus the perceived economic benefit to modern society. Issues such as stable carbon storage ultimately establish a cost and practicality bottleneck for many carbon capture processes.(1) Such issues can be resolved with the development of techniques that synergistically capture and convert atmospheric emissions into materials that can be developed into high-value products. (2) This produces a secondary market for greenhouse gas emissions and provides an economic value to pollutants that otherwise challenge the promise of long-term human sustainability on Earth.

In this manner, the elemental constituents of carbon dioxide, the most notable greenhouse gas, involve carbon and oxygen, which are foundational elemental building blocks for technological systems. Specifically, carbon-based materials are widely used in applications. One of the most notable applications of carbon is for anodes in lithium-ion (Li-ion) batteries, which are the principal rechargeable battery for electric vehicles (EVs) and consumer electronics.(3-5) Commercial Li-ion batteries most commonly rely on anodes produced with graphite that exhibit a theoretical Li-anode capacity of 1 Li: 6 C, or 372 mAh g^{-1}, (5) and an observed capacity of 280–320 mAh g^{-1}.(6) Because of the greater Earth abundance of Na compared to Li (2.3% vs lithium's 0.0017% in the Earth's crust), recent efforts have also focused on carbon-based anodes for Na-ion battery systems.(7-9) A key challenge has been the low capacity of Na in crystalline carbons (32–35 mAh g^{-1}) which can be improved by introducing defects into the lattice or engineering the electrode–electrolyte interface to facilitate solvent-assisted intercalation.(10-12) Whereas other materials besides carbon can form low-potential compounds practical for Na-ion and Li-ion anodes, such as Si and Sn, (9, 13) issues of rapid capacity fade, solid-electrolyte interphase vulnerability, (14) and existing commercial manufacturing infrastructure relevant to carbon-based anodes all present numerous technological challenges in transitioning battery systems away from carbon-based electrodes. Most recently, efforts to combine carbon-based Earth-abundant electrode materials, such as banana peels and peat moss, with sodium-ion batteries has been forward progress in this research area.(15, 16)

In this report, we build upon the solar thermal electrochemical process (STEP), (17-21) which is designed to convert greenhouse gas carbon dioxide into a useful carbon commodity. This technique uses inexpensive electrode materials (galvanized steel cathode and a nickel anode) and molten carbonate electrolytes that are heated and powered using concentrated photovoltaic (CPV) cells that convert sunlight into electricity at 39% efficiency. STEP has been shown to function effectively with or without solar powered operation to electrolytically split water, carbon dioxide, or metal oxides, (22-24) produce STEP carbon, (18) produce STEP ammonia and STEP organic, (25-27) and produce STEP iron or cement.(28-30) Here we show that this process can be used as a sustainable synthetic pathway for defect-controlled CNT and CNF materials, which exhibit excellent performance in the context of lithium-ion and sodium-ion battery anode materials. This presents a sustainable route to convert carbon dioxide into materials relevant to both grid-scale and portable storage systems.

2. RESULTS AND DISCUSSION

The technique utilized to produce CNTs and CNFs from the STEP process is illustrated in Figure1. In this study CNTs are grown by DC electrolysis from (natural isotope abundance) CO_2dissolved in 750 °C molten Li_2CO_3 with, or without, added Li_2O. A 100 mL Ni crucible serves as both container and (O_2 generating) anode, and immersed 10 cm^2 galvanized steel as the cathode. The carbon product is characterized in the bottom of Figure 1. Following an initial low current (0.001 A for 0.5 h) step to grow Ni nucleation sites on the cathode, CNTs are grown on an immersed 10 cm^2 galvanized steel cathode at 1 A for 1 h. Two types of nanostructures are generated: straight CNTs that are grown in electrolyte without added Li_2O, and tangled CNTs that are grown when 4 m Li_2O has been added to the electrolyte.

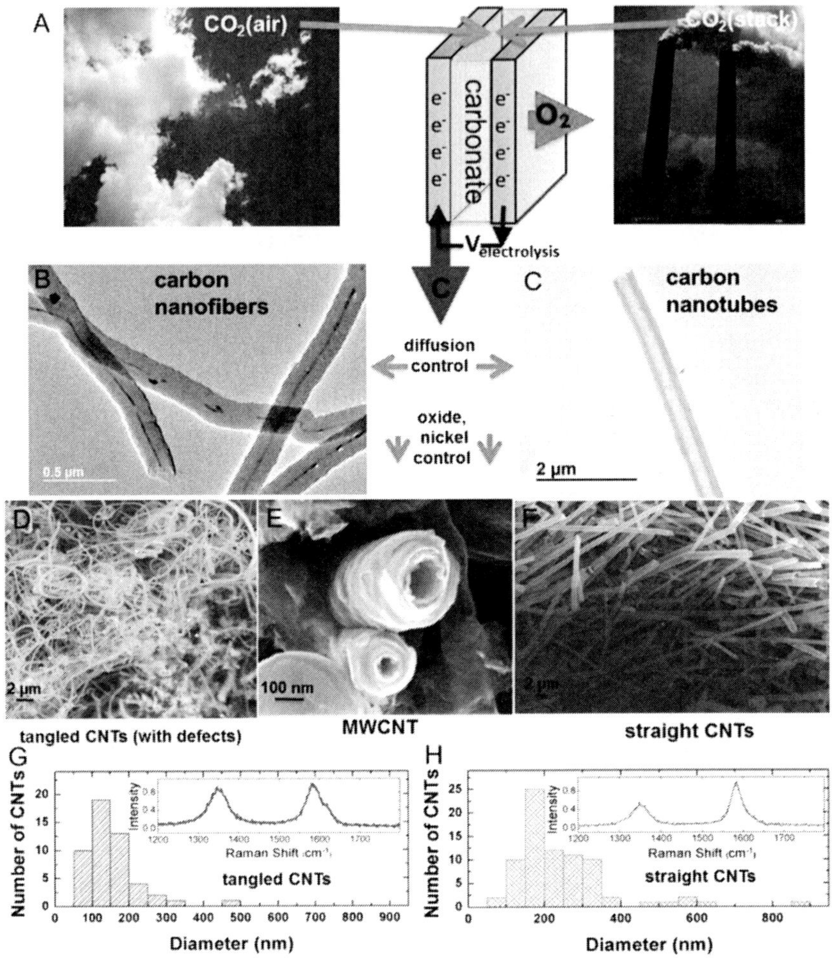

Figure 1. (A) Concept of high yield electrolytic synthesis of carbon nanostructures from dissolved air or smokestack concentrations of CO_2 in molten lithiated carbonates. During CO_2 electrolysis, transition metal deposition controls the nucleation and morphology of the carbon nanostructure. (B–C) SEM images depicting the different CNT products formed by controlled diffusion. SEM in (B) is from ^{13}C, and SEM in (C) is grown from natural abundance CO_2. (D–F) SEM images showing the different CNT morphologies formed based on either the addition of Li_2O (D - tangled, defective) or the absence of Li_2O (F - straight, less defective). (E) Edge-on high magnification view of STEP CNTs. (G, H) Diameter distribution of straight (G) and tangled (H) CNTs based on image analysis of SEM images, and Raman spectra of CNTs as used in this study and synthesized at 750 °C from natural abundance ^{12}C.

The control of diffusion conditions during electrolytic splitting of CO_2 in molten lithium carbonate leads to either filled CNF or hollow CNT nanostructures, and control of oxide and transition metal concentration leads to tangled or straight fibers. This gives a level of control on the synthesized carbon nanostructures critical for battery applications. Specifically, the $^{13}CO_2$ and $Li_2{}^{13}CO_3$ CNF (Figure 1B) and 99% ^{12}C on the CNT (Figure 1C) illustrated are grown under similar conditions, but diffusion restraints of the heavier isotope allow more frequent Ni nucleation points, which can tend to fill the interior of the tube nanostructure. Similarly, both straight CNTs (Figure 1D) and tangled CNTs (Figure 1F) can be produced. The straight CNTs shown are grown without added oxide. Using this process, larger diameter CNTs can be obtained by pulsing the formation current for 9 min on (at 1 A) and 1 min off. The tangled CNTs shown are grown in high electrolytic oxide DC conditions.(18)

Raman spectroscopic analysis was carried out to study the degree of graphitization of the synthesized carbon nanostructures. In Figure 1, the Raman spectrum exhibits two sharp peaks observed at 1350 and 1580 cm^{-1}, which correspond to the disorder-induced mode (D band) and the high frequency E_{2g} first order mode (G band) that correspond generally to sp^3 and sp^2hybridized carbon species, respectively. The intensity ratio between the D band and G band (I_D/I_G) is an important parameter to evaluate the graphitization and hence the total relative ratio of defective carbons in the material. As shown in Figure 1, the I_D/I_G ratio for tangled CNTs is significantly higher than straight CNTs—the latter of which is consistent with commercial hollow carbon nanofiber samples.(31) As we show in this study, synthetic control of the D: G ratio in a carbon material is useful to engineer the intercalation properties of carbon-based electrodes. Furthermore, based on a collection of SEM images of the as-grown tangled and straight CNTs, size distributions of the CNT materials were assigned based on Image J analysis software.(32)Size distributions indicate that the tangled CNTs exhibit a slightly overall smaller CNT diameter than the straight CNTs; however, all CNTs in this study are in a size range that is ideal for battery materials. The sizes of CNTs in this study (1) minimizes electrolyte consumption due to SEI formation in comparison to smaller nanostructured carbons and (2) enables full accessibility of the carbon material to alkali ion diffusion in thick, 3D electrode slurries.

To demonstrate the capability to transform CO_2 into a usable carbon material that can be assessed for energy storage applications, two types of CNTs, straight and tangled, were developed into electrodes, combined into half-cells along with electrolyte and a separator and pressed into coin cells for electrochemical testing. For Li-ion cells, CO_2-derived CNT electrodes were cast into a slurry with conductive carbon black and PVDF binder (3:1:1 ratio). This electrode was then combined with a separator, 1 M $LiPF_6$ in EC/DEC electrolyte, lithium metal foil, and pressed into a 2032 coin cell. Cyclic voltammetry tests (Supporting Information) and galvanostatic charge–discharge tests at rates of 100 mA/g elucidate the storage capability of both tangled (Figure 2A) and straight (Figure 2B) CO_2-derived CNTs. The first discharge, which is associated with solid electrolyte interphase (SEI) formation, is isolated from subsequent discharge cycles and plotted relative to the top axis in Figure 2A,B. For both the tangled and straight CNTs, subsequent cycling leads to high Coulombic efficiency (near 100%) and stable performance by the ~15th cycle. In both cases, the reversible capacity on the second discharge is measured near 370 mAh g^{-1}, and this stabilizes near 350 mAh g^{-1} by the 15th cycle in both cases. To further assess the long-term performance of these materials, we carried out extended cycling tests at 100 mA/g (~C/7.5) rates for 200 cycles, which extended for ~2.5 months of continuous testing (Figure 2C). Here, a distinction between the storage behavior of straight (less defective) and tangled (higher defect content) CNTs emerges. Whereas the capacity of straight CNTs remains virtually unchanged over the cycling process, the storage capacity of the tangled CNTs is observed to steadily increase. After 200 cycles, the tangled CNT capacity is measured as ~460 mAh g^{-1}, with the capacity of straight CNTs remaining invariant at ~360 mAh g^{-1}. As such increased capacity above 372 mAh g^{-1},(33, 34) and specifically during cycling,(35) has been reported in other studies on carbon nanostructured electrodes, our results imply this effect may be related to defect-induced modification to storage processes over the course of cycling. On the basis of the comparison of straight CNTs to tangled CNTs, the high defect content and torturous bends in the tangled CNTs likely could enable a transition from dilute staging of Li^+, which occurs during the formation of LiC_6,(36) to a combination of this and a mechanism analogous to pore-filling, which has been observed with other alkali-ion systems with very high defect-containing carbons.(11) The increased capacity over cycling emphasizes the continuous activation of this storage mode, which enables a capacity superior to the maximum alloying capacity of Li in LiC_6.

In addition to lithium-ion batteries, we also analyzed these materials as electrodes for Na^+ battery anodes. Recent efforts have been focused on Na^+ storage mechanisms into carbon materials,(11)and whereas some reports indicate intercalation,(15, 16) often involving solvent cointercalation,(12) defect-containing carbon materials are known to achieve moderate Na^+ storage capacities based on a combined intercalation and pore-filling mechanism that only occurs with highly defective materials.(11) In this manner, the ability to modulate the defect density based on the STEP synthesis pathway enables a comparison that can elucidate this effect and highlights a synthetic trajectory toward high performance CO_2-derived sodium-ion battery electrodes for grid-scale applications. In this case, the batteries were prepared identically, except the electrolyte was based on 1 M $NaPF_6$ in diethylene glycol dimethyl ether (DGM), and the half-cells were cycled against Na foil counterelectrodes. Unlike the case of Li-ion cells, the Na-ion cells exhibit significantly different storage performance between the straight and tangled CNTs (Figure 3). In this case, galvanostatic charge–discharge data elucidates the straight CNTs to exhibit a reversible capacity that is only slightly higher than the maximum intercalation capacity of Na^+ in crystalline carbons. Whereas the I_D/I_G ratio is ~0.4, which implies a highly defective material relative to crystalline carbon, galvanostatic data indicate that this defect density is still too low to access a significant capacity of stored Na^+ in the context of pore-filling mechanism. However, the tangled CNTs, which exhibit a higher I_D/I_G ratio near 0.9, exhibit reversible capacities over 130 mAh g^{-1}, which is ~2× that of the straight CNTs. This implies defects in a CNT material are critical to activate the mechanism for Na^+ storage, and this is achieved in the tangled CNTs produced in the STEP process. To further assess the stability of this anode performance, both straight and tangled CNTs were cycled for 600 cycles at a similar rate of 100 mA/g. In parallel to Li-ion cells, this represents ~2.5 months of continuous cycling of the devices. Over the course of this cycling process, the devices show invariant performance with no observed capacity fade, which is improved compared to other defective carbon materials that exhibit storage via the pore-filling mechanism.(37) One possible explanation for this improved performance is a storage capacity that appears to originate mostly from the sloping part of the galvanostatic Na^+ insertion curve. Recent work by Bommier et al. has proposed that this sloping region is correlated with defect-activated sodium insertion in hard carbons.(11) This is distinguished from the flat, lower-voltage feature attributed primarily to plating of Na^+ on the interior of micropores in the anode. Our results are consistent with this picture since by increasing the defect density of the CNTs, the total capacity of the sloping region in the galvanostatic curves also similarly increases, but notably the flat signature at low voltages remains generally absent. Therefore, our results not only are in agreement with Bommier et al., but the invariant cycling performance observed over 600 cycles (~2.5 months) implies that defect-activated storage is highly reversible and not as prone to the capacity fade observed when sodium insertion occurs primarily through the plating mechanism. This implies that high capacity sodium-ion batteries with excellent cycling performance can be rationally designed by controlling structural and defect properties of the carbons. Results from cyclic voltammetry scans and rate capability tests of these materials are available in the Supporting Information.

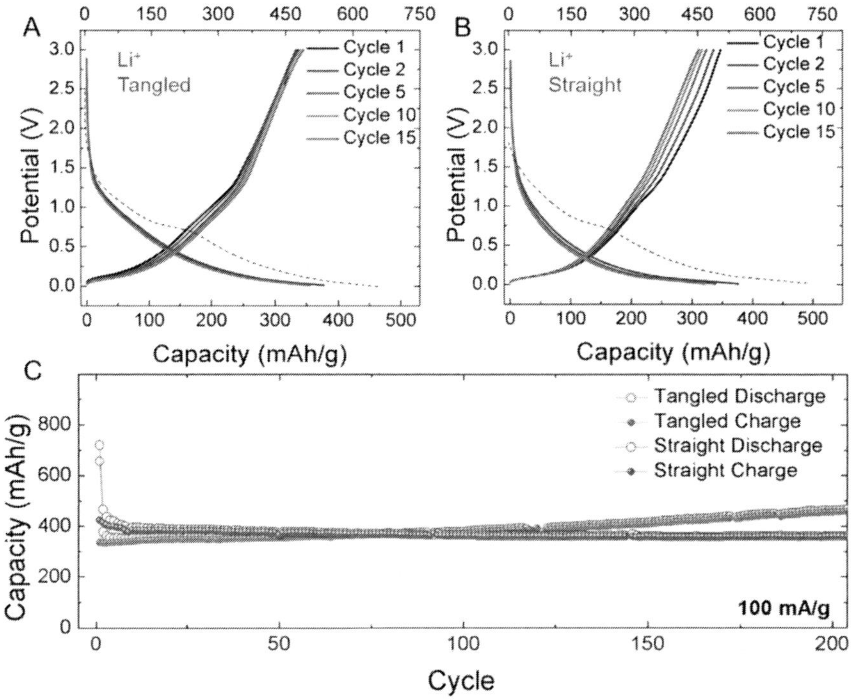

Figure 2. CO_2-derived Li-ion batteries. (A, B) First 15 galvanostatic charge–discharge profiles for CO_2-derived straight and tangled CNTs at a current density of 100 mA/g. The first discharge (dashed line) is longer than subsequent discharge cycles due to SEI formation and is referenced to the top axis in mAh/g. (C) Extended cycling performed at a current density of 100 mAh/g over 200 cycles for both straight and tangled CNTs.

Overall, electrochemical tests give promise to the function of CO_2-derived CNTs as practical anode materials for batteries. This establishes the principle that energy input can transform CO_2, which is a global pollutant with adverse environmental impact, into a secondary product that now is associated with an economic value in a thriving technological area. In order to illustrate this point, simple calculations were performed based on available data in the literature to correlate the average value of CO_2 in batteries based upon the total cost per kWh of the battery cell (Figure 4). Unlike routes to transform CO_2 into fuels such as methanol, which targets a low-value hydrocarbon commodity, the transformation of CO_2 into active battery materials provides a valuation of the CO_2 that is associated with the total cost of the battery technology. The U.S. Department of Energy Vehicle Technology Office (FY 2015 Budget Outlook (38)) has set a target of $125 per kWh by year 2022, with an average battery cost of ~$325 per kWh in 2013. These calculations, which are very generalized and can widely vary based on battery manufacturing techniques, battery size, and packaging processes, build upon the assumptions that (1) ~8% of the total battery cost is associated with the anode,(39) (2) the total mass of anodic carbon in a 1 kWh module is <10% of the total weight of the battery, (3) the mass extrapolated for a 1 kWh module is ~5.3 kg (e.g., Panasonic 18650 type 1.5 kWh modules), and (4) processes to convert CO_2 to functional carbon materials transform all carbon into usable material. This explicitly demonstrates an operation window for large-scale processes capable of converting CO_2 into carbon-based battery anodes and explicitly demonstrates the principle that CO_2 can be associated with economic value in a growing technological sector.

In this spirit, two possibilities emerge to produce materials derived from CO_2 that can have economic value in battery systems: ambient atmospheric capture of CO_2 or the extraction of $CO_2$from industrial smokestacks, such as from conventional fossil fuel power plants. The energy costs of the latter based on combined cycle (CC) natural gas plants are discussed in brief. Rather than processing large volumes to transform CO_2 from air, the CO_2 is available at higher concentration, and not only provides the hot CO_2 for dissolution in the electrolysis, but also removes CO_2 from the plant emission. Additionally, the electrolysis provides a pure oxygen product for improved power plant combustion efficiency. Figure 5A illustrates action of a conventional CC electrical power plant which emits a flue gas that contains ~9% CO_2. The conventional CC plant increases the fuel to electricity efficiency compared to single cycle electrical power plants by directing the exhaust heat emissions from a gas (Brayton cycle) electrical turbine to boil water to power a steam (Rankine cycle) electrical turbine.

As shown in Figure 5B, an alternative CNT CC plant product provides a platform for the simplified CNT cost analysis, and in addition to electricity, simultaneously produces a valuable carbon nanofiber product. Unlike the conventional CC (Figure 5A), CO_2 is removed from the exhaust of the CNT CC power plant on the right. As shown, the hot CO_2, N_2, and H_2O CC emission is instead bubbled into molten carbonate where only the CO_2 dissolves. The remaining gases (with the CO_2removed) become the exhaust gas (after heat recovery). The dissolved CO_2 is split by electrolysis into oxygen gas at the anode, which is fed (after heat recovery) back into the gas turbine and carbon (CNT) at the cathode. The CNT product can hence be tuned for strength, diameter, length, defect-content, geometry, and electrical or thermal conductivity by the specific molten salt electrochemistry employed. The CNT product may be removed periodically or as a constant throughput). The recovered heat boils water to power a steam turbine to also generate electricity. Heat is returned to the steam generator chamber using (i) heat recovered from the electrolysis (pure oxygen and carbon nanofiber) products, and (ii) from the carbon dioxide removed electrolysis exhaust.

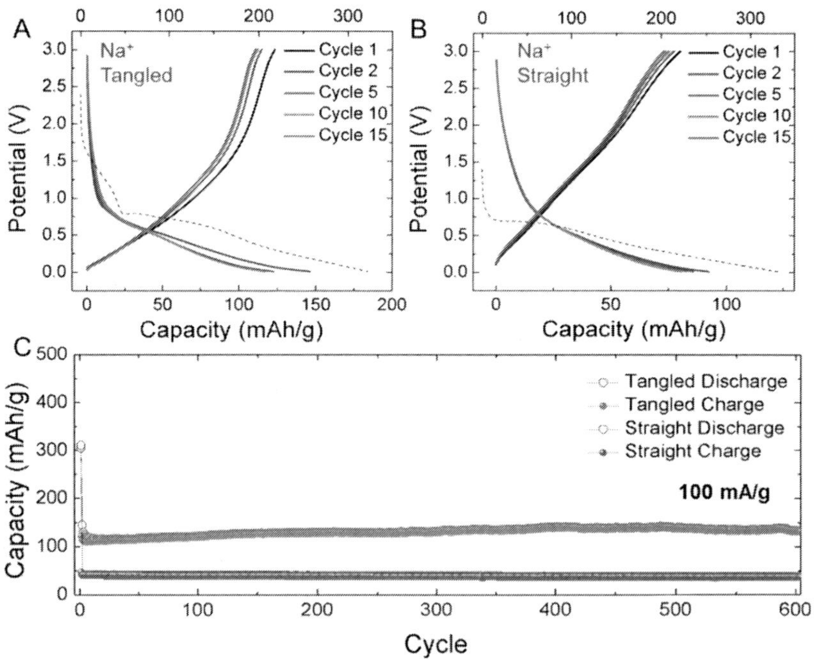

Figure 3. CO_2-derived Na-ion batteries. (A–B) First 15 galvanostatic charge–discharge profiles for CO_2-derived straight and tangled CNTs at current density of 100 mA/g. The first discharge (dashed line) is longer than subsequent discharge cycles due to SEI formation and is referenced to the top axis in mAh/g. (C) Extended cycling performed at current density of 100 mAh/g over 600 cycles for both straight and tangled CNTs.

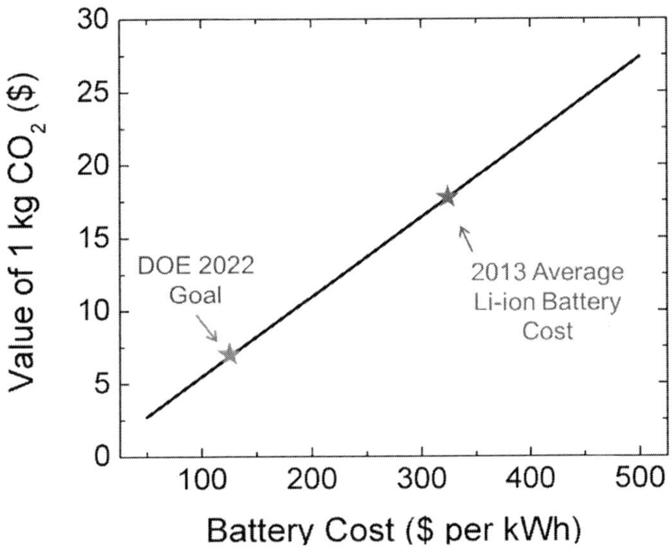

Figure 4. Generalized value of 1 kg CO_2 converted into CNT materials for batteries based on the total cost per kWh for the battery. The DOE target of $125/kWh for 2022 and the 2013 average Li-ion battery cost provide a window ranging from ~$5 to $18 of secondary value per kg of CO_2. Anode cost relative to the full cell is extrapolated from data in ref 39, and weight per kWh is extrapolated from a Panasonic 1.5 kWh (52 Ah) module.

The CNT CC natural gas to electricity plant efficiency is increased by the "free" pure oxygen generated during the CNT production by electrolysis by increasing the natural gas combustion temperature compared to the more dilute oxygen in air and contributes to an offset of the energy required to drive the electrolysis. The heated CO_2 reactant is provided by the plant, and the electrical energy required to drive the electrolysis of CO_2 is recovered many fold by the increased value of the CNT product (e.g., Figure 5) compared to the cost of the natural gas consumed. Electrolysis costs to produce CNTs will be similar to infrastructure costs associated with the chlor-alkali and aluminum industries. The electrical energy costs to prepare the CNFs are low, requiring 0.9 to 1.4 V.(21) CNFs are consistently prepared here at 80 to 100% Coulombic efficiency of the four electrons required to reduce CO_2, which at $0.10 per kWh is equivalent to $800 to $1,600 per metric tonne of CNT—leading to high revenue windows for battery applications. Lithium carbonate is not consumed during the CO_2 electrolysis and at today's cost of $6,000 per ton, as amortized over 10 year's usage, cost an additional $140 per metric tonne CNT. These costs compare to today's cost of $200,000 to 400,000 per metric tonne of industrial grade (90% purity) CNTs. The production of CNTs by electrolysis provides a low-cost pathway for CNTs and hence gives an economically practical trajectory toward the conversion of CO_2 into battery materials.

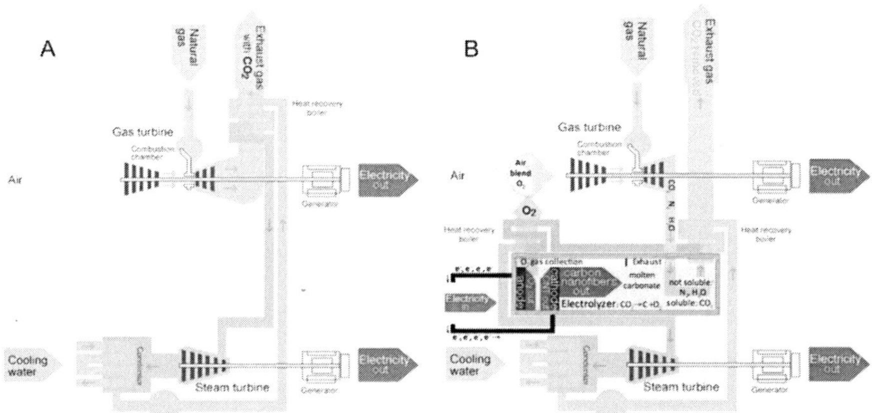

Figure 5. Action of a conventional CC power plant which has an exhaust with CO_2 (A) compared to the introduced CNF CC power plant with the carbon dioxide removed from the exhaust gas (B).(40) Left (A) The conventional gas/steam combined cycle CC power plant is illustrated as modified from ref 41. Right (B) The CC CNF power plant described in this study, illustrating the CNF (including carbon nanofibers or carbon nanotube) product, the electrolysis pure oxygen cycled back into the gas combustion, recovered heat to the steam turbine and carbon dioxide removed from the exhaust gas.

3. CONCLUSIONS

Here we report the transformation of CO_2 into low-defect (straight) and higher defect (tangled) CNT materials for use in both lithium-ion and sodium-ion batteries using the STEP process. These battery materials show excellent performance and durability, with no capacity fade measured in over 2.5 months of continuous cycling, corresponding to over 200 cycles and 600 cycles for lithium-ion and sodium-ion devices, respectively. Control on the defect density was observed to be critical to enable capacities that surpass LiC_6 in Li-ion cells and overcome the bottleneck of 30–40 mAh/g capacity in sodium-ion cells that is associated with carbon materials. This provides a bridge toward associating economic value to CO_2, with a revenue window controlled by the cost of conventional battery technology per kWh. Processes to scale this STEP technique to levels of industrial CO_2-containing smokestacks give promise for high energy efficiency CNT production through the STEP process with a STEP byproduct of O_2 that can aid combustion.

ACKNOWLEDGMENT

We thank Rizia Bardhan for use of the Raman microscope critical for this work. S.L. is grateful for partial support of this research by NSF Grants Nos. 1230732 and 1505830. A.D. is supported by the National Science Foundation Graduate Research Fellowship under Grant No. 1445197.

REFERENCES

1. Leung, D. Y. C.; Caramanna, G.; Maroto-Valer, M. M.An Overview of Current Status of Carbon Dioxide Capture and Storage Technologies *Renewable Sustainable Energy Rev.* **2014**, 39, 426– 443

2. Lim, X. Z.How to Make the Most of Carbon Dioxode *Nature* **2015**, 526, 628– 630

3. Dunn, B.; Kamath, H.; Tarascon, J. M.Electrical Energy Storage for the Grid: A Battery of Choices *Science***2011**, 334, 928– 935

4. Tarascon, J. M.Is lithium the new gold? *Nat. Chem.* **2010**, 2, 510– 510

5. Nitta, N.; Wu, F. X.; Lee, J. T.; Yushin, G.Li-ion battery materials: present and future *Mater. Today* **2015**, 18,252– 264

6. Xiong, Z.; Yun, Y. S.; Jin, H. J.Applications of Carbon Nanotubes for Lithium Ion Battery Anodes *Materials***2013**, 6, 1138– 1158

7. Kundu, D.; Talaie, E.; Duffort, V.; Nazar, L. F.The Emerging Chemistry of Sodium Ion Batteries for Electrochemical Energy Storage *Angew. Chem., Int. Ed.* **2015**, 54, 3431– 3448

8. Palomares, V.; Casas-Cabanas, M.; Castillo-Martinez, E.; Han, M. H.; Rojo, T.Update on Na-based battery materials. A growing research path *Energy Environ. Sci.* **2013**, 6, 2312– 2337

9. Yabuuchi, N.; Kubota, K.; Dahbi, M.; Komaba, S.Research development on sodium-ion batteries *Chem. Rev.***2014**, 114, 11636–11682

10. Zhu, Z.; Cheng, F.; Hu, Z.; Niu, Z.; Chen, J.Highly stable and ultrafast electrode reaction of graphite for sodium ion batteries *J. Power Sources* **2015**, 293, 626– 634

11. Bommier, C.; Surta, T. W.; Dolgos, M.; Ji, X.New Mechanistic Insights on Na-Ion Storage in Nongraphitizable Carbon *Nano Lett.* **2015**, 15, 5888– 5892

12. Cohn, A. P.; Share, K.; Carter, R.; Oakes, L.; Pint, C. L.Ultrafast Solvent-Assisted Sodium Ion Intercalation into Highly Crystalline Few-Layered Graphene *Nano Lett.* **2016**, 16, 543– 548

13. Wu, Q. L.; Li, J. C.; Xiao, X. C.; Lott, A.; Lu, W. Q.; Sheldon, B. W.; Wu, J.Silicon-Based Nanomaterials for Lithium-Ion Batteries: A Review *Adv. Energy Mater.* **2014**, 4, 1300882

14. Menkin, S.; Golodnitsky, D.; Peled, E.Artificial solid-electrolyte interphase (SEI) for improved cycleability and safety of lithium-ion cells for EV applications *Electrochem. Commun.* **2009**, 11, 1789–1791

15. Ding, J.; Wang, H. L.; Li, Z.; Kohandehghan, A.; Cui, K.; Xu, Z. W.; Zahiri, B.; Tan, X. H.; Lotfabad, E. M.;Olsen, B. C.Carbon Nanosheet Frameworks Derived from Peat Moss as High Performance Sodium Ion Battery Anodes *ACS Nano* **2013**, 7, 11004– 11015

16. Lotfabad, E. M.; Ding, J.; Cui, K.; Kohandehghan, A.; Kalisvaart, W. P.; Hazelton, M.; Mitlin, D.High-Density Sodium and Lithium Ion Battery Anodes from Banana Peels *ACS Nano* **2014**, 8, 7115– 7129

17. Licht, S.STEP (Solar Thermal Electrochemical Photo) Generation of Energetic Molecules: A Solar Chemical Process to End

Anthropogenic Global Warming *J. Phys. Chem. C* **2009**, 113, 16283–16292

18. Ren, J. W.; Li, F. F.; Lau, J.; Gonzalez-Urbina, L.; Licht, S.One-Pot Synthesis of Carbon Nanofibers from CO2 *Nano Lett.* **2015**, 15, 6142– 6148

19. Licht, S.; Wang, B. H.; Wu, H. J.STEP-A Solar Chemical Process to End Anthropogenic Global Warming. II: Experimental Results *J. Phys. Chem. C* **2011**, 115, 11803– 11821

20. Licht, S.; Wang, B. H.; Ghosh, S.; Ayub, H.; Jiang, D. L.; Ganley, J.A New Solar Carbon Capture Process: Solar Thermal Electrochemical Photo (STEP) Carbon Capture *J. Phys. Chem. Lett.* **2010**, 1, 2363– 2368

21. Ren, J. W.; Lau, J.; Lefler, M.; Licht, S.The Minimum Electrolytic Energy Needed To Convert Carbon Dioxide to Carbon by Electrolysis in Carbonate Melts *J. Phys. Chem. C* **2015**, 119, 23342–23349

22. Licht, S.Efficient Solar-Driven Synthesis, Carbon Capture, and Desalinization, STEP: Solar Thermal Electrochemical Production of Fuels, Metals, Bleach *Adv. Mater.* **2011**, 23, 5592– 5612

23. Li, F. F.; Liu, S. Z.; Cui, B. C.; Lau, J.; Stuart, J.; Wang, B. H.; Licht, S.Solar Fuels: A One-Pot Synthesis of Hydrogen and Carbon Fuels from Water and Carbon Dioxide *Adv. Energy Mater.* **2015**, 5, 1401791

24. Licht, S.; Halperin, L.; Kalina, M.; Zidman, M.; Halperin, N.Electrochemical potential tuned solar water splitting *Chem. Commun.* **2003**, 3006– 3007

25. Licht, S.; Cui, B. C.; Wang, B. H.; Li, F. F.; Lau, J.; Liu, S. Z.Ammonia synthesis by N-2 and steam electrolysis in molten hydroxide suspensions of nanoscale Fe2O3 *Science* **2014**, 345, 637– 640

26. F.; Licht, S.Advances in Understanding the Mechanism and Improved Stability of the Synthesis of Ammonia from Air and Water in Hydroxide Suspensions of Nanoscale Fe2O3 *Inorg. Chem.* **2014**, 53,10042– 10044

27. Zhu, Y. J.; Wang, B. H.; Liu, X. L.; Wang, H. Y.; Wu, H. J.; Licht, S.STEP organic synthesis: an efficient solar, electrochemical process for the synthesis of benzoic acid *Green Chem.* **2014**, 16, 4758– 4766

28. Cui, B. C.; Licht, S.Critical STEP advances for sustainable iron production *Green Chem.* **2013**, 15, 881–884

29. Licht, S.; Wu, H. J.STEP Iron, a Chemistry of Iron Formation without CO_2 Emission: Molten Carbonate Solubility and Electrochemistry of Iron Ore Impurities *J. Phys. Chem. C* **2011**, 115, 25138– 25147

30. Licht, S.; Wu, H. J.; Hettige, C.; Wang, B. H.; Asercion, J.; Lau, J.; Stuart, J.STEP cement: Solar Thermal Electrochemical Production of CaO without CO_2 emission *Chem. Commun.* **2012**, 48, 6019– 6021,

31. Saravanan, M.; Sennu, P.; Ganesan, M.; Ambalavanan, S.Multi-Walled Carbon Nanotubes Percolation Network Enhanced the Performance of Negative Electrode for Lead-Acid Battery *J. Electrochem. Soc.* **2013**,160, A70– A76,

32. Schneider, C. A.; Rasband, W. S.; Eliceiri, K. W.NIH Image to ImageJ: 25 years of image analysis *Nat. Methods* **2012**, 9, 671– 675

33. Lahiri, I.; Oh, S. W.; Hwang, J. Y.; Cho, S.; Sun, Y. K.; Banerjee, R.; Choi, W.High Capacity and Excellent Stability of Lithium Ion Battery Anode Using Interface-Controlled Binder-Free Multiwall Carbon Nanotubes Grown on Copper *ACS Nano* **2010**, 4, 3440– 3446

34. Carter, R.; Oakes, L.; Cohn, A. P.; Holzgrafe, J.; Zarick, H. F.; Chatterjee, S.; Bardhan, R.; Pint, C. L.Solution Assembled Single-Walled Carbon Nanotube Foams: Superior Performance in Supercapacitors, Lithium-Ion, and Lithium-Air Batteries *J. Phys. Chem. C* **2014**, 118, 20137– 20151

35. Fan, Z. J.; Yan, J.; Ning, G. Q.; Wei, T.; Zhi, L. J.; Wei, F.Porous graphene networks as high performance anode materials for lithium ion batteries *Carbon* **2013**, 60, 558– 561

36. Inaba, M.; Yoshida, H.; Ogumi, Z.; Abe, T.; Mizutani, Y.; Asano, M.In situ Raman study on electrochemical Li intercalation into graphite *J. Electrochem. Soc.* **1995**, 142, 20– 26

37. Cao, Y.; Xiao, L.; Sushko, M.; Wang, W.; Schwenzer, B.; Xiao, J.; Nie, Z.; Saraf, L. V.; Yang, Z.; Liu,J.Sodium ion insertion in hollow carbon nanowires for battery applications *Nano Lett.* **2012**, 12, 3783– 3787

38. Technologies Office FY 2015 Budget at-a-Glance; U.S. Department of Energy: Washington, **2014**; http://energy.gov/sites/ prod/files /2014/03/ f9/fy15 _at-a-glance_vto.pdf.

39. Wood, D. L.; Li, J. L.; Daniel, C.Prospects for reducing the processing cost of lithium ion batteries *J. Power Sources* **2015**, 275, 234– 242

40. Licht, S.Carbon Nanofiber Electric and Heat Power Plants, U.S. Provisional Patent 62/240,805, October 13, **2015**.

41. eon.co, accessed November 9, **2015**, at http://www.eon.com/en/business-areas/power-generation/natural-gas-and-oil/combined-cycle-gas-turbine-how-it-works.html

INDEX

A

Ambient Carbon Dioxide, 267
Armchair Swcnts, 29

B

Bias Voltage, 240

C

Carbon black, 223
carbon nanotubes, 1, 36, 37, 38, 224, 229, 248, 249, 250, 251, 253, 255, 256, 262, 263, 264, 265, 267
Comocat Swcnts, 77, 82, 83
Coupled RBM vibrations, 252

D

Double-walled Boron Nitride Nanotubes, 115

E

electron diffraction, 42, 43, 101, 102, 112, 250, 251, 252, 253, 262, 263

F

full batteries, 208, 209, 217

L

lithium half cells, 216

M

mesoporous nanotubes, 199, 200, 202, 203, 204, 205, 206, 207, 208, 209, 210, 211, 212, 213, 214, 215, 217
Morphology, 98, 216

N

nickel-based composite catalyst, 52

P

pea-like nanotubes, 199, 200, 202, 204, 205, 206, 207, 208, 209, 212, 215, 216
photon energy, 169, 171, 172, 174, 175, 178, 189, 190, 253, 257, 258, 262
Polyacetylene, 85

R

Raman spectra, 63

S

sodium batteries, 217
sonication, 52, 180, 231, 234, 235, 237, 238
supercapacitors, 199, 201, 205, 210, 212, 217, 222

T

Test fixture, 236
tunable mean diameter, 108

U

up-conversion light emission, 171, 173

Z

Zigzag Swcnts, 27